# Sustainable Energy

# Sustainable Energy

Annette Stone

MURPHY & MOORE
www.murphy-moorepublishing.com

Published by Murphy & Moore Publishing,
1 Rockefeller Plaza,
New York City, NY 10020, USA
www.murphy-moorepublishing.com

**Sustainable Energy**
Annette Stone



International Standard Book Number: 978-1-63987-512-2 (Hardback)



**Cataloging-in-Publication Data**

Sustainable energy / Annette Stone.
p. cm.
Includes bibliographical references and index.
ISBN 978-1-63987-512-2
1. Renewable energy sources. 2. Sustainable development. I. Stone, Annette.
TJ808 .S87 2022
333.794--dc23

# TABLE OF CONTENTS

# PREFACE

The purpose of this book is to help students understand the fundamental concepts of this discipline. It is designed to motivate students to learn and prosper. I am grateful for the support of my colleagues. I would also like to acknowledge the encouragement of my family.

The practice of using energy in a sustained way that meets the current demand of energy without compromising the needs of future generations is known as sustainable energy. This type of energy is renewable. It is helpful in reducing greenhouse gas emissions and causes no damage to the environment. Solar energy is the energy that is obtained from the sun and is harnessed using a wide range of technologies such as photovoltaics, solar heating, solar thermal energy, etc. Wind energy is a type of sustainable energy which uses wind to provide mechanical power using wind turbines. Hydropower is the generation of electricity from the energy of falling or fast running water. Geothermal energy refers to the thermal energy which is generated and stored in the Earth's crust. This book unfolds the innovative aspects of sustainable energy which will be crucial for the holistic understanding of the subject matter. Sustainable energy is an upcoming field of science that has undergone rapid development over the past few decades. This book will serve as a reference to a broad spectrum of readers.

A foreword for all the chapters is provided below:

Chapter - What is Sustainable Energy?

Sustainable energy refers to the practice of using and conserving energy in such a way that it meets the demands and needs of the present. Clean energy management, renewable and non-renewable resources, usage of sustainable transport, etc. fall under its domain. The topics elaborated in this chapter will help in gaining a better perspective of sustainable energy.

Chapter - Wind Energy

Wind energy is the sustainable energy that uses wind for generating mechanical power through wind turbines. Some types of wind turbines are airborne wind turbines, floating wind turbines, vertical axis wind turbines, Savonius wind turbines, etc. This chapter sheds light on the subject of wind energy to provide an in-depth understanding of it.

Chapter - Solar Energy

Solar energy is the radiant light and heat from the sun that is utilized for generating electricity or causing chemical reactions. It use a range of technologies such as solar heating, solar thermal energy, photovoltaics, solar architecture, etc. This chapter closely examines solar energy to provide an extensive understanding of the subject.

Chapter - Hydropower

The power that is derived from the energy of falling or fast-running water is referred to as hydropower. Small hydro, micro hydro, pico hydro, marine current power, tidal stream generator, etc. are a few of its concepts. This chapter delves into the subject of hydropower for a thorough understanding of it.

Chapter - Bioenergy

Bioenergy is a form of renewable energy that generates electricity from organic matter. Biofuels, biodiesel, biogasoline, biorefinery, etc. come under the subject of bioenergy. All the aspects related to bioenergy have been carefully analyzed in this chapter.

**Annette Stone**

# What is Sustainable Energy?

# 1

- Energy
- Sustainable Energy
- Clean Energy and its Management

Sustainable energy refers to the practice of using and conserving energy in such a way that it meets the demands and needs of the present. Clean energy management, renewable and non-renewable resources, usage of sustainable transport, etc. fall under its domain. The topics elaborated in this chapter will help in gaining a better perspective of sustainable energy.

## Energy

Energy, in physics means the capacity for doing work. It may exist in potential, kinetic, thermal, electrical, chemical, nuclear, or other various forms. There are, moreover, heat and work–i.e., energy in the process of transfer from one body to another. After it has been transferred, energy is always designated according to its nature. Hence, heat transferred may become thermal energy, while work done may manifest itself in the form of mechanical energy.

All forms of energy are associated with motion. For example, any given body has kinetic energy if it is in motion. A tensioned device such as a bow or spring, though at rest, has the potential for creating motion; it contains potential energy because of its configuration. Similarly, nuclear energy is potential energy because it results from the configuration of subatomic particles in the nucleus of an atom.

Energy can be neither created nor destroyed but only changed from one form to another. This principle is known as the conservation of energy or the first law of thermodynamics. For example, when a box slides down a hill, the potential energy that the box has from being located high up on the slope is converted to kinetic energy, energy of motion. As the box slows to a stop through friction, the kinetic energy from the box's motion is converted to thermal energy that heats the box and the slope.

Energy can be converted from one form to another in various other ways. Usable mechanical or electrical energy is, for instance, produced by many kinds of devices, including fuel-burning heat engines, generators, batteries, fuel cells, and magnetohydrodynamic systems.

In the International System of Units (SI), energy is measured in joules. One joule is equal to the work done by a one-newton force acting over a one-metre distance.

## Sustainable Energy

Sustainable energy is a form of energy that meet our today's demand of energy without putting them in danger of getting expired or depleted and can be used over and over again. Sustainable energy should be widely encouraged as it do not cause any harm to the environment and is available widely free of cost. All renewable energy sources like solar, wind, geothermal, hydropower and ocean energy are sustainable as they are stable and available in plenty.

Sun will continue to provide sunlight till we all are here on earth, heat caused by sun will continue to produce winds, earth will continue to produce heat from inside and will not cool down anytime soon, movement of earth, sun and moon will not stop and this will keep on producing tides and the process of evaporation will cause water to evaporate that will fall down in the form of rain or ice which will go through rivers or streams and merge in the oceans and can be used to produce energy through hydropower. This clearly states that all these renewable energy sources are sustainable and will continue to provide energy to the coming generations.

There are many forms of sustainable energy sources that can be incorporated by countries to stop the use of fossil fuels. Sustainable energy does not include any sources that are derived from fossil fuels or waste products. This energy is replenishable and helps us to reduce greenhouse gas emissions and causes no damage to the environment. If we are going to use fossil fuels at a steady rate, they will expire soon and cause adverse affect to our planet.

Fossil fuels are not considered as sustainable energy sources because they are limited, cause immense pollution by releasing harmful gases and are not available everywhere on earth. Fossil fuels normally include coal, oil and natural gas. Steps must be taken to reduce our dependency on fossil fuels as pose dangerous to environment. Most of the counties have already started taking steps to make use of alternative energy sources. As of today, around 20% of world's energy needs comes from renewable energy sources. Hydropower is the most common form of alternative energy used around the world.

## Need for Sustainable Energy

During ancient times, wood, timber and waste products were the only major energy sources. In short, biomass was the only way to get energy. When more technology was developed, fossil fuels like coal, oil and natural gas were discovered. Fossil fuels proved boom to the mankind as they were widely available and could be harnessed easily. When these fossil fuels were started using extensively by all the countries across the globe, they led to degradation of environment. Coal and oil are two of the major sources that produce large amount of carbon dioxide in the air. This led to increase in global warming.

Also, few countries have hold on these valuable products which led to the rise in prices of these fuels. Now, with rising prices, increasing air pollution and risk of getting expired soon forced scientists to look out for some alternative or renewable energy sources. The need of the hour was to look for resources that are available widely, cause no pollution and are replenishable. Sustainable Energy, at that time came into the picture as it could meet our today's increasing demand of energy and also provide us with an option to make use of them in future also.

## Types of Sustainable Energy

Sustainable energy are not just a part of renewable energy sources, they are also the sources of energy that can best be used to power homes and industries without any harmful effects being experienced. This is the sole reason why many people advice the use of these forms of energy in everyday life. It is because its effects to the environment are purely beneficial.

## Solar Energy

Solar energy is the best form of sustainable energy. This energy manifests itself in tow forms. There is the light and the heat. Both of these forms are equally important to us in our day to day living and other forms of life. For instance, the plants need the light to grow and generate food while man needs the heat energy to maintain body temperature and power their homes and industries. This means that it is the greatest form of sustainable energy. It can be used two folds with greater results as needed. This only serves to generate confidence and ensure that we live the way we intended without causing further harm to the environment.

According to activists, it is the future of energy. Evidence of intensive use of this alternative energy source can be seen everywhere. There are many companies that are making solar panels to tap this energy for use at home or in the industries. Consequently, the energy is also being tapped for commercial purposes in many fields like powering of homes in power grids. All that one needs to do is to get hold of the solar panel and install it in the homes or commercial property. During the summer periods, you can cut down on your energy costs.

## Wind Energy

Wind is a sustainable energy source. It is available naturally and can be tapped to produce vast amounts of power that can be used in many ways and places. For instance, sailors tap this energy to help the ship propel through its various directions to distant shores for trading. Nowadays, this energy sources is being commercialized. There are many companies that have invested heavily on power grids and windmills to tap into this energy source. The energy generated can be sold to other people to power their homes and industries. In the near future, sustainable energy like wind power will be a big industry and the fossil fuels exploration will have halted and no longer being used.

## Geothermal Energy

Geothermal energy allows us fetch the energy from beneath the earth. This occurs by installing geothermal power stations that can use heat coming out from inside the earth and use it to generate electricity. The temperature below the earth around 10,000 meters is so high that it can used to boil water. Geothermal energy cannot be harnessed everywhere as high temperature is needed to produce steam that could move turbines. It can be harnessed in those areas that have high seismic activity and are prone to volcanoes. They are environment friendly and can produce energy throughout the day but their ability to produce energy at suitable regions restricts us from using it on a much wider scale.

### Ocean Energy

There is massive size of oceans in this world. About 70% of the earth is covered with water. The potential that ocean energy has to generate power is much higher than any other source of energy. This sustainable energy allows us to harness it in 3 ways i.e. wave, tidal or ocean thermal energy conversion (OTEC). Tides have immense power which when effectively tapped can generate a lot of energy and can be used to power millions of homes. Waves produced at the oceans can be used by ocean thermal plants to convert the kinetic energy in waves to mechanical energy of turbines which can again converted to electrical energy through generators. Setting up of big plants at ocean may cause ecological imbalance and disturb aquatic life.

### Biomass Energy

Biomass energy is produced by burning of wood, timber, landfills and municipal and agricultural waste. It is completely renewable and does not produce harmful gases like carbon dioxide which is primarily responsible for increase in global warming. Although, carbon dioxide is produced by burning these products but that is equally compensated when plants take this carbon dioxide and produce oxygen. It also helps to reduce landfills but are not as effective as fossil fuels.

### Hydroelectric Power

On the other hand, there are the rivers or waterfalls whose energy of the moving water is captured that can turn turbines to generate power. This is commonly known as hydroelectric power. It is very common nowadays and it is powering most parts of the world and one of the biggest form of alternative energy currently being used. There are many companies and countries that are exporting this energy to other countries who unable to harness it on their own due to lack of the necessary resources or conditions. The energy is commonly transported in form of power lines to various parts of the country and even outside the country.

These are the three best case examples of sustainable energy forms that are projected to run the world in the near future. They are very sustainable and so not cause any environmental effects. Their inability to be depleted and lack of effect to the environmental makes them a perfect candidate to future energy needs.

## Clean Energy and its Management

Energy is at the heart of most critical economic, environmental and developmental issues facing the world today. Clean, efficient, affordable and reliable energy services are indispensable for global prosperity. Energy management and optimization solution can help reduce energy costs while improving mill operational performance.

Access to reliable, affordable and sustainable energy is essential for improving living standards, development and economic growth. To overcome poverty and improve people health in developing country, it is essential to expand access to reliable and clean energy. In this way, they will be able to increase productivity and promoting economic growth. Challenges such as fuel shortages, high energy costs, global warming and environmental issues must drive policies that target more affordable and sustainable energy solutions. In essence, one way to overcome poverty, promote health and educational services and enhance socioeconomic development is to ensure reliable, sustainable and affordable energy for everyone.

The ratio between achieved performance or the profits from services, goods or energy, and the energy used to achieve this is energy efficiency and based on DIN 4602, energy management is the predictive, organized and systematic coordination of the procurement, conversion, distribution and use of energy to cover requirements while taking account of ecological and economic aims. The term thus describes actions for the purpose of efficient energy handling. Energy management and optimization solutions can help reduce energy costs. Energy use of buildings presents 40% of the total primary energy consumption in the United States. Often, this energy is consumed inefficiently. Most of the problems predominantly arise from building technical operation and energy management. A study on commercial buildings has found that 50% of the buildings have control problems. Savings of up to 77% have been achieved by correction of control problems. For optimization energy use, there should be a planning and scheduling. Real-time data form monitoring systems and production planning are necessary for scheduling. This information coupled with price and availability information from energy markets used to optimize power generation plant. Significant challenges arise from the increasing connection of variable renewable energy generation. Despite their long-term benefits in terms of sustainability, renewable resources such as solar and wind powers are not only highly variable but also partially unpredictable. They, therefore, present operational challenges in assessing and mitigating risk, including issues of imbalance between generation and demand and network constraints. Additionally, there is longer term uncertainty over the technology costs and environmental policies associated with renewable generation, among other factors. These uncertainties at various time scales may imply the need for significant investment in new, more flexible 'smart grid' technologies capable of adaptation to a range of future scenarios. As laid out in, examples are:

- Controllable distributed energy assets including storage.
- More flexible transmission systems.
- Corrective control following network outages.
- Wide area monitoring and control equipment and communications.
- Controllable demand and the active participation of end users in, for example, system balancing and congestion management. Beside all software and methodology to manage energy consumption, ISO 50001 is a sustainable business tool that

helps organizations implements a flexible and robust energy management system (EMS). Effective energy management is not just good for business; it is also becoming a requirement. ISO 50001 will help organization understand how they are using various types of energy and identify realistic ways of reducing consumption, emissions and costs. The international standard outlines energy management practices that not only save the organization money today, but also in the long term; all while helping shield the bottom line from the increasing cost of energy. ISO 50001 also shows the commitment to reducing environmental impact which can help the businesses stand out from their competition. For implementing ISO 50001, first, management responsibility should be determined. After that, energy review should be done to identify baselines. Performance indicators will be used to evaluate how successfully the EMS is operating. In the next step, specific guidance on what needs to be communicated to whom as an EMS is being planned, implemented, maintained or improved. Documentation is essential for EMS and supporting information such as energy consumption bills. Non-conformities are identified via the audit process as the non-fulfillment of a requirement of the standard; corrective actions are what the actions an organization must take in order to fulfill the requirement. Finally, management review should be done to evaluate the progress and achievements of the EMS.

## References


- Energy, science: britannica.com, Retrieved 02 July, 2020
- Sustainableenergy: conserve-energy-future.com, Retrieved 28 May, 2020
- Various-renewable-energy-sources: conserve-energy-future.com, Retrieved 16 April, 2020
- Clean-energy-management, energy-management-for-sustainable-development: intechopen.com, Retrieved 19 August, 2020

# Wind Energy 2

- Wind Power Forecasting
- Wind Resource Assesment
- Wind Turbines
- Airborne Wind Turbine
- Floating Wind Turbine
- Vertical Axis Wind Turbine
- Savonius Wind Turbine
- Small Wind Turbine
- Unconventional Wind Turbines
- Darrieus Wind Turbine

Wind energy is the sustainable energy that uses wind for generating mechanical power through wind turbines. Some types of wind turbines are airborne wind turbines, floating wind turbines, vertical axis wind turbines, Savonius wind turbines, etc. This chapter sheds light on the subject of wind energy to provide an in-depth understanding of it.

Wind Energy is the most mature and developed renewable energy Wind energy (or wind power) describes the process by which wind is used to generate electricity. Wind turbines convert the kinetic energy in the wind into mechanical power. A generator can convert mechanical power into electricity. Mechanical power can also be utilized directly for specific tasks such as pumping water. The US DOE developed a short wind power animation that provides an overview of how a wind turbine works and describes the wind resources in the United States.

Wind is caused by the uneven heating of the atmosphere by the sun, variations in the earth's surface, and rotation of the earth. Mountains, bodies of water, and vegetation all influence wind flow patterns. Wind turbines convert the energy in wind to electricity by rotating propeller-like blades around a rotor. The rotor turns the drive shaft, which turns an electric generator. Three key factors affect the amount of energy a turbine can harness from the wind: wind speed, air density, and swept area.

Equation for wind power:

$$P = \frac{1}{2}\rho A V^3$$

## Wind Speed

The amount of energy in the wind varies with the cube of the wind speed, in other words, if the wind speed doubles, there is eight times more energy in the wind ($2^3 = 2 \times 2 \times 2 = 8$). Small changes in wind speed have a large impact on the amount of power available in the wind.

## Density of the Air

The more dense the air, the more energy received by the turbine. Air density varies with elevation and temperature. Air is less dense at higher elevations than at sea level, and warm air is less dense than cold air. All else being equal, turbines will produce more power at lower elevations and in locations with cooler average temperature.

## Swept Area of the Turbine

The larger the swept area (the size of the area through which the rotor spins), the more power the turbine can capture from the wind. Since swept area is $A = pir^2$, where r = radius of the rotor, a small increase in blade length results in a larger increase in the power available to the turbine.

## Wind Farm Development

Siting a wind farm varies from one location to another, but there are some important matters for land owners to consider:

- Understand your wind resource.
- Evaluate distance from existing transmission lines.
- Determine benefits of and barriers to allowing your land to be developed.
- Establish access to capital.
- Identify reliable power purchaser or market.
- Address siting and project feasibility considerations.
- Understand wind energy's economics.
- Obtain zoning and permitting expertise.
- Establish dialogue with turbine manufacturers and project developers.
- Secure agreement to meet O and M needs.

## Land Requirements

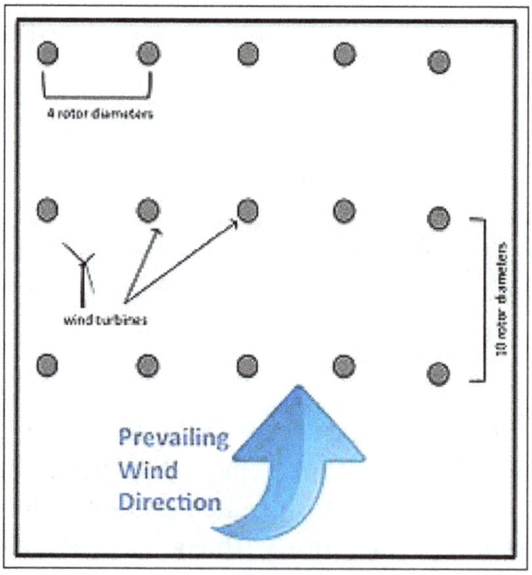


The amount of land required for a wind farm varies considerably, and is particularly dependent on two key factors: the desired size of the wind farm (which can be defined either by installed capacity or the number of turbines) and the characteristics of the local terrain. Typically, wind turbine spacing is determined by the rotor diameter and local wind conditions. Some estimates suggest spacing turbines between 5 and 10 rotor diameters apart. If prevailing winds are generally from the same direction, turbines may be installed 3 or 4 rotor diameters apart (in the direction perpendicular to the prevailing winds); under multi-directional wind conditions, spacing of between 5 and 7 rotor diameters is recommended.

## Wind Power Forecasting

A wind power forecast corresponds to an estimate of the expected production of one or more wind turbines (referred to as a wind farm) in the near future. By production is often meant available power for wind farm considered (with units kW or MW depending on the wind farm nominal capacity). Forecasts can also be expressed in terms of energy, by integrating power production over each time interval.

### Time Scales of Forecasts

Forecasting of the wind power generation may be considered at different time scales, depending on the intended application. From milliseconds up to a few minutes, forecasts can be used for the turbine active control. Such type of forecasts are usually referred to as very short-term forecasts. For the following 48–72 hours, forecasts are needed for the power system management or energy trading. They may serve for deciding on the use of conventional power plants (Unit commitment) and for the optimization of the scheduling of these plants (Economic dispatch). Bids for energy to be supplied on a day are usually required during the morning of the previous day. These forecasts are called short-term forecasts. For longer time scales (up to 5–7 days ahead), forecasts may be considered for planning the maintenance of wind farms, or conventional power plants or transmission lines. Maintenance of offshore wind farms may be particularly costly, so optimal planning of maintenance operations is of particular importance.

For the last two possibilities, the temporal resolution of wind power predictions ranges between 10 minutes and a few hours (depending on the forecast length). Improvements of wind power forecasting has focused on using more data as input to the models involved, and on providing uncertainty estimates along with the traditionally provided predictions.

### Reason for Wind Power Forecasts

In the electricity grid at any moment balance must be maintained between electricity consumption and generation – otherwise disturbances in power quality or supply may occur. Wind generation is a direct function of wind speed and, in contrast to conventional generation systems, is not easily dispatchable. Fluctuations of wind generation thus receive a great amount of attention. Variability of wind generation can be regarded at various time scales. First, wind power production is subject to seasonal variations, i.e. it may be higher in winter in Northern Europe due to low-pressure meteorological systems or it may be higher in summer in the Mediterranean regions owing to strong summer breezes. There are also daily cycles which may be substantial, mainly due to daily temperature changes. Finally, fluctuations are observed at the very short-term scale (at the minute or intra-minute scale). The variations are not of the same order for these three different timescales. Managing the variability of wind generation is the key aspect associated to the optimal integration of that renewable energy into electricity grids.

The challenges to face when wind generation is injected in a power system depend on the share of that renewable energy. It is a basic concept, the wind penetration which allows one to describe the share of wind generation in the electricity mix of a given power system. For Denmark, which is a country with one of the highest shares of wind power in the electricity mix, the average wind power penetration over the year is of 16–20% (meaning that 16–20% of the electricity consumption is met wind energy), while the instantaneous penetration (that is, the instantaneous wind power production compared to the consumption to be met at a given time) may be above 100%.

The transmission system operator (TSO) is responsible for managing the electricity balance on the grid: at any time, electricity production has to match consumption. Therefore, the use of production means is scheduled in advance in order to respond to load profiles. The load corresponds to the total electricity consumption over the area of interest. Load profiles are usually given by load forecasts which are of high accuracy. For making up the daily schedule, TSOs may consider their own power production means, if they have any, and they can purchase power generation from Independent Power Producers (IPPs) and utilities, via bilateral contracts or electricity pools. In the context of deregulation, more and more players appear on the market, thus breaking the traditional situation of vertically-integrated utilities with quasi local monopolies. Two main mechanisms compose electricity markets. The first one is the spot market where participants propose quantities of energy for the following day at a given production cost. An auction system permits to settle the electricity spot price for the various periods depending on the different bids. The second mechanism is the balancing of power generation, which is coordinated by the TSO. Depending on the energy lacks and surplus (e.g. due to power plant failures or to intermittence in the case of wind power installations), the TSO determines the penalties that will be paid by IPPs who missed in their obligations. In some cases, an intra-day market is also present, in order to take corrective actions.

In order to illustrate this electricity market mechanism, consider the Dutch electricity market. Market participants, referred to as Program Responsible Parties (PRPs), submit their price-quantity bids before 11 am for the delivery period covering the following day from midnight to midnight. The Program Time Unit (PTU) on the balancing market is of 15 minutes. Balancing of the 15-minute averaged power is required from all electrical producers and consumers connected to the grid, who for this purpose may be organised in sub-sets. Since these sub-sets are referred to as Programmes, balancing on the 15-minute scale is referred to as Programme Balance. Programme Balance now is maintained by using the production schedules issued the day before delivery and measurement reports (distributed the day after delivery). When the measured power is not equal to the scheduled power, the Programme Imbalance is the difference between the realised sum of production and consumption and the forecast sum of production and consumption. If only production from wind energy is taken into account, Programme Imbalance reduces to realised wind production minus forecast wind production. The programme imbalance is the wind production forecast error.

Programme Imbalance is settled by the System Operator, with different tariffs for negative Programme Imbalance and positive Programme Imbalance. A positive Programme Imbalance indicates more energy actually produced than forecast. by wind energy the realised wind production is bigger than the forecast wind production. And vice versa, in the case of a negative Programme Imbalance by wind energy.

Note that the costs for positive and negative imbalances may be asymmetric, depending on the balancing market mechanism. In general, wind power producers are penalized by such market system since a great part of their production may be subject to penalties.

In parallel to be used for market participation, wind power forecasts may be used for the optimal combined operation of wind and conventional generation, wind and hydro-power generation, or wind in combination with some energy storage devices. They also serve as a basis for quantifying the reserve needs for compensating the eventual lacks of wind production.

## General Methodology

Several methods are used for short-term prediction of wind generation. The simplest ones are based on climatology or averages of past production values. They may be considered as reference forecasting methods since they are easy to implement, as well as benchmark when evaluating more advanced approaches. The most popular of these reference methods is certainly persistence. This naive predictor – commonly referred to as 'what you see is what you get' — states that the future wind generation will be the same as the last measured value. Despite its apparent simplicity, this naive method might be hard to beat for look-ahead times up to 4–6 hours ahead.

Advanced approaches for short-term wind power forecasting necessitate predictions of meteorological variables as input. Then, they differ in the way predictions of meteorological variables are converted to predictions of wind power production, through the so-called power curve. Such advanced methods are traditionally divided into two groups. The first group, referred to as physical approach, focuses on the description of the wind flow around and inside the wind farm, and use the manufacturer's power curve, for proposing an estimation of the wind power output. In parallel the second group, referred to as statistical approach, concentrates on capturing the relation between meteorological predictions (and possibly historical measurements) and power output through statistical models whose parameters have to be estimated from data, without making any assumption on the physical phenomena.

## Prediction of Meteorological Variables

Wind power generation is directly linked to weather conditions and thus the first aspect of wind power forecasting is the prediction of future values of the necessary weather variables at the level of the wind farm. This is done by using numerical weather prediction (NWP) models. Such models are based on equations governing the motions and forces affecting motion of fluids. From the knowledge of the actual state of the atmosphere, the system of equations allows to estimate what the evolution of state variables, e.g. temperature, velocity, humidity and pressure, will be at a series of grid points. The meteorological variables that are needed as input for wind power prediction obviously include wind speed and direction, but also possibly temperature, pressure and humidity. The distance between grid points is called the spatial resolution of the NWPs. The mesh typically has spacing that varies between few kilometers and up to 50 kilometers for mesoscale models. Regarding the time axis, the forecast length of most of the operational models today is between 48 and 172 hours ahead, which is in adequacy with the requirements for the wind power application. The temporal resolution is usually between 1 and 3 hours. NWP models impose their temporal resolution to short-term wind power forecasting methods since they are used as a direct input.

Predictions of meteorological variables are provided by meteorological institutes. Meteorologists employ atmospheric models for weather forecasts on short and medium term periods. An atmospheric model is a numerical approximation of the physical description of the state of the atmosphere in the near future, and usually is run on a supercomputer. Each computation starts with initial conditions originating from recent measurements. The output consists of the expected instantaneous value of physical quantities at various vertical levels in a horizontal grid and stepping in time up to several hours after initiation. There are several reasons why atmospheric models only approximate reality. First of all, not all relevant atmospheric processes are included in the model. Also, the initial conditions may contain errors (which in a worse case propagate), and the output is only available for discrete points in space (horizontal as well as vertical) and time. Finally, the initial conditions age with time – they are already old when the computation starts let alone when the output is published. Predictions of meteorological variables are issued several times per day (commonly between 2 and 4 times per day), and are available few hours after the beginning of the forecast period. This is because some time is needed for acquiring and analyzing the wealth of measurements used as input to NWP models, then run the model and check and distribute the output forecast series. This gap is a blind spot in the forecasts from an atmospheric model. As an example in the Netherlands, KNMI publishes 4 times per day expected values of wind speed, wind direction, temperature and pressure for the period the between 0 and 48 hours after initialization of the atmospheric model Hirlam with measured data, and then the period before forecast delivery is of 4 hours.

Many different atmospheric models are available, ranging from academic research tools to fully operational instruments. Besides for the very nature of the model (physical processes or numerical schemes) there are some clear distinctive differences between them: time domain (from several hours to 6 days ahead), area (several 10.000 $km^2$ to an area covering half the planet), horizontal resolution (1 km to 100 km) and temporal resolution (1 hour to several hours).

One of the atmospheric models is the High Resolution Limited Area Model, abbreviated HiRLAM, which is frequently used in Europe. HiRLAM comes in many versions; that is why it is better to speak about “a” HiRLAM rather than “the” HiRLAM. Each version is maintained by a national institute such as the Dutch KNMI, the Danish DMI or Finnish FMI. And each institute has several versions under her wing, divided into categories such as: operational, pre-operational, semi operational and for research purposes.

Other atmospheric models are UKMO in the UK, Lokalmodell in Germany, Alladin in France (Alladin and Lokalmodell are also used by other countries within Europe), and MM5 in the USA.

## Physical Approach to Wind Power Forecasting

Meteorological forecasts are given at specific nodes of a grid covering an area. Since wind farms are not situated on these nodes, it is then needed to extrapolate these forecasts at the desired location and at turbine hub height. Physical-based forecasting methods consist of several sub-models which altogether deliver the translation from the wind forecast at a certain grid point and model level, to power forecast at the site considered. Every sub-model contains the mathematical description of the physical processes relevant to the translation. Knowledge of all relevant processes is therefore crucial when developing a purely physical prediction method (such as the early versions of the Danish Prediktor). The core idea of physical approaches is to refine the NWPs by using physical considerations about the terrain such as the roughness, orography and obstacles, and by modeling the local wind profile possibly accounting for atmospheric stability. The two main alternatives to do so are: (i) to combine the modeling of the wind profile (with a logarithmic assumption in most of the cases) and the geostrophic drag law for obtaining surface winds; (ii) to use a CFD (Computational Fluid Dynamics) code that allows one to accurately compute the wind field that the farm will see, considering a full description of the terrain.

When the wind at the level of the wind farm and at hub height is known, the second step consists in converting wind speed to power. Usually, that task is carried out with theoretical power curves provided by the wind turbine manufacturer. However, since several studies have shown the interest of using empirically derived power curve instead of theoretical ones, theoretical power curves are less and less considered. When applying a physical methodology, the modeling of the function which gives the wind generation from NWPs at given locations around the wind farm is done once for all. Then, the estimated transfer function is consequently applied to the available weather predictions at a given moment. In order to account for systematic forecasting errors that may be due to the NWP model or to their modeling approach, physical modelers often integrate Model Output Statistics (MOS) for post-processing power forecasts.

## Statistical Approach to Wind Power Forecasting

Statistical prediction methods are based on one or several models that establish the relation between historical values of power, as well as historical and forecast values of meteorological variables, and wind power measurements. The physical phenomena are not decomposed and accounted for, even if expertise of the problem is crucial for choosing the right meteorological variables and designing suitable models. Model parameters are estimated from a set of past available data, and they are regularly updated during online operation by accounting for any newly available information (i.e. meteorological forecasts and power measurements).

Statistical models include linear and non-linear models, but also structural and black-box types of models. Structural models rely on the analyst's expertise on the phenomenon of interest while black-box models require little subject-matter knowledge and are constructed from data in a fairly mechanical way. Concerning wind power forecasting, structural models would be those that include a modeling of the diurnal wind speed variations, or an explicit function of meteorological variable predictions. Black-box models include most of the artificial-intelligence-based models such as Neural-Networks (NNs) and Support Vector Machines (SVMs). However, some models are 'in-between' the two extremes of being completely black-box or structural. This is the case of expert systems, which learn from experience (from a dataset), and for which prior knowledge can be injected. We then talk about grey-box modeling. Statistical models are usually composed by an autoregressive part, for seizing the persistent behavior of the wind, and by a 'meteorological' part, which consists in the nonlinear transformation of meteorological variable forecasts. The autoregressive part permits to significantly enhance forecast accuracy for horizons up to 6–10 hours ahead, i.e. over a period during which the sole use of meteorological forecast information may not be sufficient for outperforming persistence.

Today, major developments of statistical approaches to wind power prediction concentrate on the use of multiple meteorological forecasts (from different meteorological offices) as input and forecast combination, as well as on the optimal use of spatially distributed measurement data for prediction error correction, or alternatively for issuing warnings on potentially large uncertainty.

## Uncertainty of Wind Power Forecasts

Current designs are optimal only for nonturbulent, steady conditions. Design tools accounting for unsteadiness and turbulence are far less developed.

Predictions of wind power output are traditionally provided in the form of point forecasts, i.e. a single value for each look-ahead time, which corresponds to the expectation or most-likely outcome. They have the advantage of being easily understandable because this single value is expected to tell everything about future power generation. Today, a major part of the research efforts on wind power forecasting still focuses on point prediction only, with the aim of assimilating more and more observations in the models or refining the resolution of physical models for better representing wind fields at the very local scale for instance. These efforts may lead to a significant decrease of the level of prediction error.

However, even by better understanding and modeling both the meteorological and power conversion processes, there will always be an inherent and irreducible uncertainty in every prediction. This epistemic uncertainty corresponds to the incomplete knowledge one has of the processes that influence future events. Therefore, in complement to point forecasts of wind generation for the coming hours or days, of major importance is to provide means for assessing online the accuracy of these predictions. In practice today, uncertainty is expressed in the form of probabilistic forecasts or with risk indices provided along with the traditional point predictions. It has been shown that some decisions related to wind power management and trading are more optimal when accounting for prediction uncertainty. For the example of the trading application, studies have shown that reliable estimation of prediction uncertainty allows wind power producer to significantly increase their income in comparison to the sole use of an advanced point forecasting method. Other studies of this type deal with optimal dynamic quantification of reserve requirements, optimal operation of combined systems including wind, or multi-area multi-stage regulation. More and more research efforts are expected on prediction uncertainty and related topics.

There are a number of questions that have still yet to be answered, according to a report from a coalition of researchers from universities, industry, and government, supported by the Atkinson Center for a Sustainable Future at Cornell University. They include:

- How do wind farms with their multiple wakes interact with the atmospheric boundary layer to determine the net power that can be produced?
- How do uneven terrain, roughness of the land or sea surface, and turbulence above the boundary layer and turbine wakes affect unsteady loading of downstream wind turbine blades?
- What is the effect of the atmospheric stability (convective, neutral, or stably stratified) on the performance and loading characteristics throughout a typical daily cycle?
- What is the optimal placement of wind turbines in an array, so that the kinetic energy capture can be maximized and unsteady loading be minimized?

The report also provides possible tools used to support this necessary research.

### Accuracy

The correlation between wind output and prediction can be relatively high, with an average uncorrected error of 8.8% in Germany over a two-year period.

## Wind Resource Assessment

Wind resource assessment is the process by which wind power developers estimate the future energy production of a wind farm. Accurate wind resource assessments are crucial to the successful development of wind farms.

### Wind Resource Maps

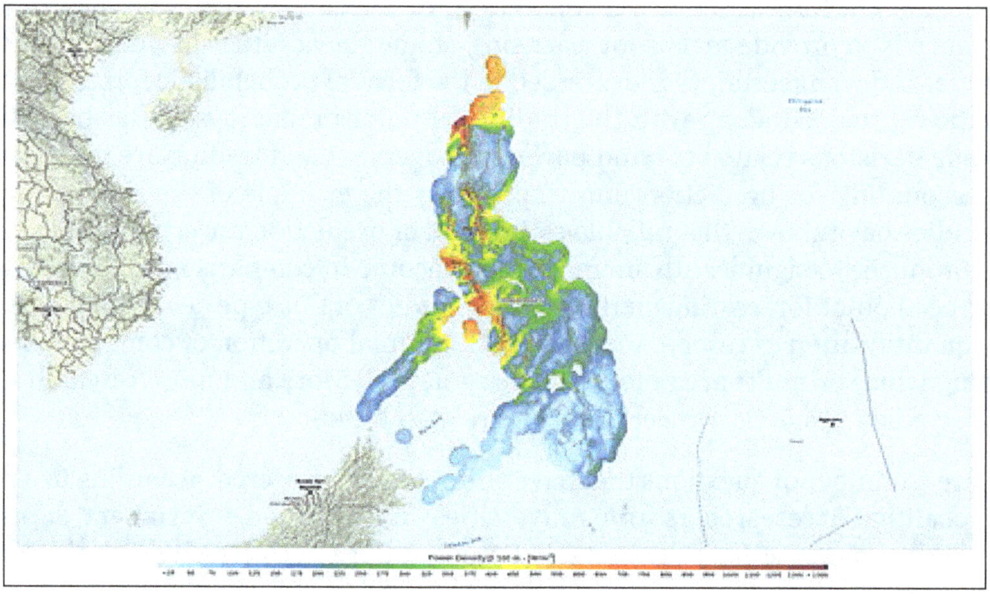

Wind resource map.

High resolution mapping of wind power resource potential has traditionally been carried out at the country level by government or research agencies, in part due to the complexity of the process and the intensive computing requirements involved. However, in 2015 the Technical University of Denmark, under framework of the Clean Energy Ministerial, launched the Global Wind Atlas (version 1.0) to provide freely available data on wind resource potential globally. The Global Wind Atlas was relaunched in November 2017 (version 2.0) in partnership with the World Bank, with wind resource maps now available for all countries at 250m resolution.

Another similar international example is the European Wind Atlas, which is in the process of being updated under the New European Wind Atlas project funded by the European Union.

Examples of country wind resource maps include the Canadian Wind Atlas, the Wind Resource Atlas of the United States, and a series of wind maps published by the World Bank under an initiative launched by ESMAP in 2013 focused on developing countries. This followed a previous initiative of the United Nations Environment Program, the Solar and Wind Energy Resource Assessment (SWERA) project, which was launched in 2002 with funding from the Global Environment Facility. However, these country wind resource maps have been largely superseded by the Global Wind Atlas in terms of data quality, methodology, and output resolution.

The above global and country mapping outputs, and many others, are also available via the Global Atlas for Renewable Energy developed by the International Renewable Energy Agency (IRENA), which brings together publicly available GIS data on wind and other renewable energy resources effort.

Wind prospecting can begin with the use of such maps, but the lack of accuracy and fine detail make them useful only for preliminary selection of sites for collecting wind speed data. With increasing numbers of ground-based measurements from specially installed anemometer stations, as well as operating data from commissioned wind farms, the accuracy of wind resource maps in many countries has improved over time, although coverage in most developing countries is still patchy. In addition to the publicly available sources listed above, maps are available as commercial products through specialist consultancies, or users of GIS software can make their own using publicly available GIS data such as the US National Renewable Energy Laboratory's High Resolution Wind Data Set.

Although the accuracy has improved, it is unlikely that wind resource maps, whether public or commercial, will eliminate the need for on-site measurements for utility-scale wind generation projects. However, mapping can help speed up the process of site identification and the existence of high quality, ground-based data can shorten the amount of time that on-site measurements need to be collected.

In addition to 'static' wind resource atlases which average estimates of wind speed and power density across multiple years, tools such as Renewables.ninja provide time-varying simulations of wind speed and power output from different wind turbine models at an hourly resolution.

## Measurements

To estimate the energy production of a wind farm, developers must first measure the wind on site. Meteorological towers equipped with anemometers, wind vanes, and sometimes temperature, pressure, and relative humidity sensors are installed. Data from these towers must be recorded for at least one year to calculate an annually representative wind speed frequency distribution.

Since onsite measurements are usually only available for a short period, data is also collected from nearby long-term reference stations (usually at airports). This data is used to adjust the onsite measured data so that the mean wind speeds are representative of a long-term period for which onsite measurements are not available. Versions of these maps can be seen and used with software applications such as Wind Navigator.

## Calculations

The following calculations are needed to accurately estimate the energy production of a proposed wind farm project:

- Correlations between onsite meteorological towers:
    - Multiple meteorological towers are usually installed on large wind farm sites. For each tower, there will be periods of time where data is missing but has been recorded at another onsite tower. Least squares linear regressions and other, more wind-specific regression methods can be used to fill in the missing data. These correlations are more accurate if the towers are located near each other (a few km distance), the sensors on the different towers are of the same type, and are mounted at the same height above the ground.
- Correlations between long term weather stations and onsite meteorological towers:
    - Because wind is variable year to year, and power produced is related to the cube of windspeed, short-term (< 5 years) onsite measurements can result in highly inaccurate energy estimates. Therefore, wind speed data from nearby longer term weather stations (usually located at airports) are used to adjust the onsite data. Least squares linear regressions are usually used, although several other methods exist as well.
- Vertical shear to extrapolate measured wind speeds to turbine hub height:
    - The hub heights of modern wind turbines are usually 80 m or greater, but developers are often reluctant to install towers taller than 60m due to the need for FAA permitting in the US, and costs. The power law and log law vertical shear profiles are the most common methods of extrapolating measured wind speed to hub height.
- Wind flow modeling to extrapolate wind speeds across a site:
    - Wind speeds can vary considerably across a wind farm site if the terrain is complex (hilly) or there are changes in roughness (the height of vegetation or buildings). Wind flow modeling software, based on either the traditional WAsP linear approach or the newer CFD approach, is used to calculate these variations in wind speed.
- Energy production using a wind turbine manufacturer's power curve:
    - When the long term hub height wind speeds have been calculated, the

manufacturer's power curve is used to calculate the gross electrical energy production of each turbine in the wind farm.

- Application of energy loss factors:
    - To calculate the net energy production of a wind farm, the following loss factors are applied to the gross energy production:
        - Wind turbine wake loss.
        - Wind turbine availability.
        - Electrical losses.
        - Blade degradation from ice/dirt/insects.
        - High/low temperature shutdown.
        - High wind speed shutdown.
        - Curtailments due to grid issues.

## Software Applications

Wind power developers use various types of software applications to assess wind resources.

## Wind Data Management

Wind data management software assists the user in gathering, storing, retrieving, analyzing, and validating wind data. Typically the wind data sets are collected directly from a data logger, located at a meteorological monitoring site, and are imported into a database. Once the data set is in the database it can be analyzed and validated using tools built into the system or it can be exported for use in external wind data analysis software, wind flow modeling software, or wind farm modeling software.

Many data logger manufacturers offer wind data management software that is compatible with their logger. These software packages will typically only gather, store, and analyze data from the manufacturer's own loggers.

Third party data management software and services exist that can accept data from a wide variety of loggers and offer more comprehensive analysis tools and data validation.

## Wind Data Analysis

Wind data analysis software assist the user in removing measurement errors from wind data sets and perform specialized statistical analysis.

## Atmospheric Simulation Modeling

Wind flow modeling methods calculate very high-resolution maps of wind flow, often at horizontal resolution finer than 100-m. When doing fine resolution modeling, to avoid exceeding available computing resource, the typical model domains used by these small-scale models have a few kilometers in the horizontal direction and several hundred meters in the vertical direction. Models with such a small domain are not capable of capturing meso-scale atmospheric phenomena that often drive wind patterns. To over come this limitation nested modeling is sometimes used.

## Wind Flow Modeling

Wind flow modeling software aims to predict important characteristics of the wind resource at locations where measurements are not available. The most commonly used such software application is WAsP, created at Risø National Laboratory in Denmark. WAsP uses a potential flow model to predict how wind flows over the terrain at a site. Meteodyn WT and WindStation are similar applications that use computational fluid dynamics (CFD) calculations instead, which are potentially more accurate, particularly for complex terrains.

## Wind Farm Modeling

Wind farm modeling software aims to simulate the behavior of a proposed or existing wind farm, most importantly to calculate its energy production. The user can usually input wind data, height and roughness contour lines, wind turbine specifications, background maps, and define objects that represent environmental restrictions. This information is then used to design a wind farm that maximizes energy production while taking restrictions and construction issues into account. There are several wind farm modeling software applications available, including ZephyCFD, Meteodyn WT, Openwind, Windfarmer, WindPRO, WindSim, and WAsP.

## Medium Scale Wind Farm Modelling

In recent years a new breed of wind farm development has grown from the increased need for distributed generation of electricity from local wind resources. This type of wind projects is mostly driven by land owners with high energetic requirements such as farmers and industrial site managers. A particular requirement from a wind modelling point of view is the inclusion of all local features such as trees, hedges and buildings as turbine hub-heights range from as little as 10m to 50m. Wind modelling approaches need to include these features but very few of the available wind modelling commercial software provide this capability. Several work groups have been set up around the world to look into this modelling requirement and companies including Digital Engineering Ltd (UK), NREL (USA), DTU Wind Energy (Denmark) are at the forefront of development in this area and look at the application of meso-CFD wind modelling techniques for this purpose.

## Wind Turbines

A wind turbine, or alternatively referred to as a wind energy converter, is a device that converts the wind's kinetic energy into electrical energy.

Thorntonbank Wind Farm, using 5 MW turbines REpower 5M.

Wind turbines are manufactured in a wide range of vertical and horizontal axis. The smallest turbines are used for applications such as battery charging for auxiliary power for boats or caravans or to power traffic warning signs. Larger turbines can be used for making contributions to a domestic power supply while selling unused power back to the utility supplier via the electrical grid. Arrays of large turbines, known as wind farms, are becoming an increasingly important source of intermittent renewable energy and are used by many countries as part of a strategy to reduce their reliance on fossil fuels. One assessment claimed that, as of 2009, wind had the "lowest relative greenhouse gas emissions, the least water consumption demands and the most favourable social impacts" compared to photovoltaic, hydro, geothermal, coal and gas.

### Efficiency

Conservation of mass requires that the amount of air entering and exiting a turbine must be equal. Accordingly, Betz's law gives the maximal achievable extraction of wind power by a wind turbine as 16/27 (59.3%) of the total kinetic energy of the air flowing through the turbine.

The maximum theoretical power output of a wind machine is thus 16/27 times the kinetic energy of the air passing through the effective disk area of the machine. If the effective area of the disk is A, and the wind velocity v, the maximum theoretical power output P is:

$$P = \frac{16}{27}\frac{1}{2}\rho v^3 A = \frac{8}{27}\rho v^3 A$$

where ρ is the air density.

Wind-to-rotor efficiency (including rotor blade friction and drag) are among the factors affecting the final price of wind power. Further inefficiencies, such as gearbox losses, generator and converter losses, reduce the power delivered by a wind turbine. To protect components from undue wear, extracted power is held constant above the rated operating speed as theoretical power increases at the cube of wind speed, further reducing theoretical efficiency. In 2001, commercial utility-connected turbines delivered 75% to 80% of the Betz limit of power extractable from the wind, at rated operating speed.

Efficiency can decrease slightly over time, one of the main reasons being dust and insect carcasses on the blades which alters the aerodynamic profile and essentially reduces the lift to drag ratio of the airfoil. Analysis of 3128 wind turbines older than 10 years in Denmark showed that half of the turbines had no decrease, while the other half saw a production decrease of 1.2% per year. Ice accretion on turbine blades has also been found to greatly reduce the efficiency of wind turbines, which is a common challenge in cold climates where in-cloud icing and freezing rain events occur. Vertical turbine designs have much lower efficiency than standard horizontal designs.

In general, more stable and constant weather conditions (most notably wind speed) result in an average of 15% greater efficiency than that of a wind turbine in unstable weather conditions, thus allowing up to a 7% increase in wind speed under stable conditions. This is due to a faster recovery wake and greater flow entrainment that occur in conditions of higher atmospheric stability. However, wind turbine wakes have been found to recover faster under unstable atmospheric conditions as opposed to a stable environment.

Different materials have been found to have varying effects on the efficiency of wind turbines. In an Ege University experiment, three wind turbines (Each with three blades with diameters of one meter) were constructed with blades made of different materials: A glass and glass/carbon epoxy, glass/carbon, and glass/polyester. When tested, the results showed that the materials with higher overall masses had a greater friction moment and thus a lower power coefficient.

## Types

Wind turbines can rotate about either a horizontal or a vertical axis, the former being both older and more common. They can also include blades, or be bladeless. Vertical designs produce less power and are less common.

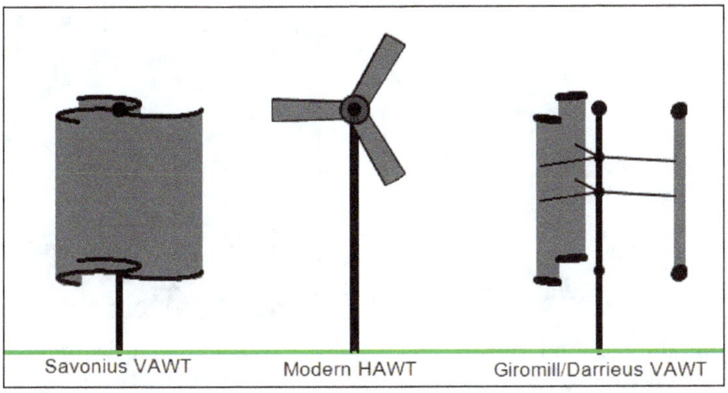


The three primary types: VAWT Savonius, HAWT towered; VAWT Darrieus as they appear in operation.

## Horizontal Axis

Components of a horizontal axis wind turbine (gearbox, rotor shaft and brake assembly) being lifted into position.

Large three-bladed Horizontal axis wind turbines (HAWT) with the blades upwind of the tower produce the overwhelming majority of wind power in the world today. These turbines have the main rotor shaft and electrical generator at the top of a tower, and must be pointed into the wind. Small turbines are pointed by a simple wind vane, while large turbines generally use a wind sensor coupled with a yaw system. Most have a gearbox, which turns the slow rotation of the blades into a quicker rotation that is more suitable to drive an electrical generator. Some turbines use a different type of generator suited to slower rotational speed input. These don't need a gearbox and are called direct-drive, meaning they couple the rotor directly to the generator with no gearbox in between. While permanent magnet direct-drive generators can be more costly due to the rare earth materials required, these gearless turbines are sometimes preferred over gearbox generators because they "eliminate the gear-speed increaser, which is susceptible to significant accumulated fatigue torque loading, related reliability issues, and maintenance costs." There is also the pseudo direct drive mechanism, which has some advantages over the permanent magnet direct drive mechanism.

Offshore Horizontal Axis Wind Turbines (HAWTs).

Onshore Horizontal Axis Wind Turbines.

Most horizontal axis turbines have their rotors upwind of the supporting tower. Downwind machines have been built, because they don't need an additional mechanism for keeping them in line with the wind. In high winds, the blades can also be allowed to bend, which reduces their swept area and thus their wind resistance. Despite these advantages, upwind designs are preferred, because the change in loading from the wind as each blade passes behind the supporting tower can cause damage to the turbine.

The rotor of a gearless wind turbine being set.

Turbines used in wind farms for commercial production of electric power are usually three-bladed. These have low torque ripple, which contributes to good reliability. The blades are usually colored white for daytime visibility by aircraft and range in length from 20 to 80 meters (66 to 262 ft). The size and height of turbines increase year by year. Offshore wind turbines are built up to 8 MW today and have a blade length up to 80 meters (260 ft). Designs with 10 to 12 MW are in preparation. Usual multi megawatt turbines have tubular steel towers with a height of 70 m to 120 m and in extremes up to 160 m.

## Vertical Axis

Vertical-axis wind turbines (or VAWTs) have the main rotor shaft arranged vertically. One advantage of this arrangement is that the turbine does not need to be pointed into the wind to be effective, which is an advantage on a site where the wind direction is highly variable. It is also an advantage when the turbine is integrated into a building because it is inherently less steerable. Also, the generator and gearbox can be placed near the ground, using a direct drive from the rotor assembly to the ground-based gearbox, improving accessibility for maintenance. However, these designs produce much less energy averaged over time, which is a major drawback.

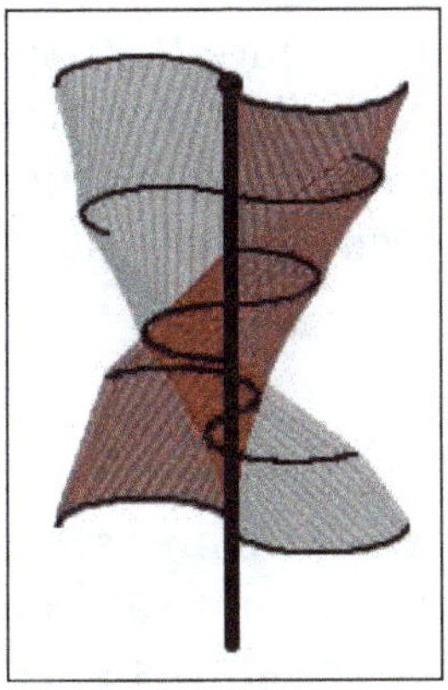

A vertical axis Twisted Savonius type turbine.

The key disadvantages include the relatively low rotational speed with the consequential higher torque and hence higher cost of the drive train, the inherently lower power coefficient, the 360-degree rotation of the aerofoil within the wind flow during each cycle and hence the highly dynamic loading on the blade, the pulsating torque generated by some rotor designs on the drive train, and the difficulty of modelling the wind flow accurately and hence the challenges of analysing and designing the rotor prior to fabricating a prototype.

When a turbine is mounted on a rooftop the building generally redirects wind over the roof and this can double the wind speed at the turbine. If the height of a rooftop mounted turbine tower is approximately 50% of the building height it is near the optimum for maximum wind energy and minimum wind turbulence. While wind speeds within the built environment are generally much lower than at exposed rural sites, noise may be a concern and an existing structure may not adequately resist the additional stress.

Subtypes of the vertical axis design include:

## Darrieus Wind Turbine

"Eggbeater" turbines, or Darrieus turbines, were named after the French inventor, Georges Darrieus. They have good efficiency, but produce large torque ripple and cyclical stress on the tower, which contributes to poor reliability. They also generally require some external power source, or an additional Savonius rotor to start turning, because the starting torque is very low. The torque ripple is reduced by using three or more blades, which results in greater solidity of the rotor. Solidity is measured by blade area divided by the rotor area. Newer Darrieus type turbines are not held up by guy-wires but have an external superstructure connected to the top bearing.

## Giromill

A subtype of Darrieus turbine with straight, as opposed to curved, blades. The cycloturbine variety has variable pitch to reduce the torque pulsation and is self-starting. The advantages of variable pitch are: high starting torque; a wide, relatively flat torque curve; a higher coefficient of performance; more efficient operation in turbulent winds; and a lower blade speed ratio which lowers blade bending stresses. Straight, V, or curved blades may be used.

## Savonius Wind Turbine

These are drag-type devices with two (or more) scoops that are used in anemometers, Flettner vents (commonly seen on bus and van roofs), and in some high-reliability low-efficiency power turbines. They are always self-starting if there are at least three scoops.

Twisted Savonius is a modified savonius, with long helical scoops to provide smooth torque. This is often used as a rooftop wind turbine and has even been adapted for ships.

## Parallel

The parallel turbine is similar to the crossflow fan or centrifugal fan. It uses the ground effect. Vertical axis turbines of this type have been tried for many years: a unit producing 10 kW was built by Israeli wind pioneer Bruce Brill in the 1980s.

## Design and Construction

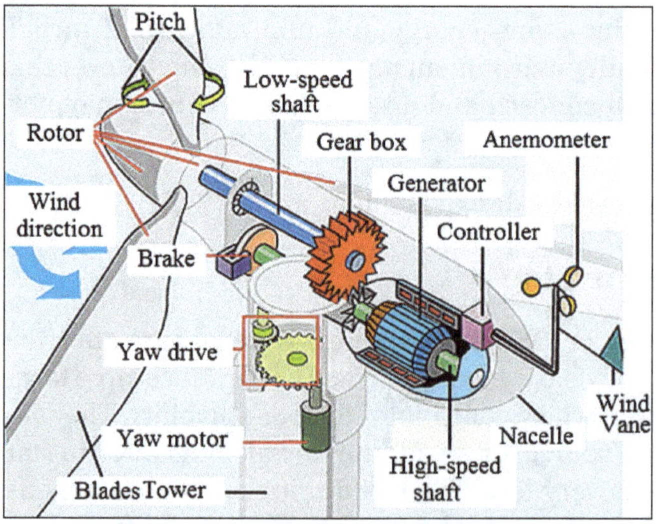


Components of a Horizontal axis wind turbine.

Inside view of a wind turbine tower, showing the tendon cables.

Wind turbine design is a careful balance of cost, energy output, and fatigue life.

## Components

Wind turbines convert wind energy to electrical energy for distribution. Conventional horizontal axis turbines can be divided into three components:

- The rotor, which is approximately 20% of the wind turbine cost, includes the blades for converting wind energy to low speed rotational energy.
- The generator, which is approximately 34% of the wind turbine cost, includes the electrical generator, the control electronics, and most likely a gearbox (e.g., planetary gear box), adjustable-speed drive, or continuously variable transmission component for converting the low-speed incoming rotation to high-speed rotation suitable for generating electricity.
- The surrounding structure, which is approximately 15% of the wind turbine cost, includes the tower and rotor yaw mechanism.

Nacelle of a wind turbine.

A 1.5 (MW) wind turbine of a type frequently seen in the United States has a tower 80 meters (260 ft) high. The rotor assembly (blades and hub) weighs 22,000 kilograms (48,000 lb). The nacelle, which contains the generator, weighs 52,000 kilograms (115,000 lb). The concrete base for the tower is constructed using 26,000 kilograms

(58,000 lb) reinforcing steel and contains 190 cubic meters (250 cu yd) of concrete. The base is 15 meters (50 ft) in diameter and 2.4 meters (8 ft) thick near the center.

## Turbine Monitoring and Diagnostics

Due to data transmission problems, structural health monitoring of wind turbines is usually performed using several accelerometers and strain gages attached to the nacelle to monitor the gearbox and equipment. Currently, digital image correlation and stereo-photogrammetry are used to measure dynamics of wind turbine blades. These methods usually measure displacement and strain to identify location of defects. Dynamic characteristics of non-rotating wind turbines have been measured using digital image correlation and photogrammetry. Three dimensional point tracking has also been used to measure rotating dynamics of wind turbines.

## Recent Developments in Technology

Wind turbine rotor blades are being made longer to increase efficiency. This requires them to be stiff, strong, light and resistant to fatigue. Materials with these properties are composites such as polyester and epoxy, while glass fiber and carbon fiber have been used for the reinforcing. Construction may use manual layup or injection molding.

## New Designs

Companies seek ways to draw greater efficiency from their designs. A predominant way has been to increase blade length and thus rotor diameter. Retrofitting existing turbines with larger blades reduces the work and risks of redesigning the system. The current longest blade is 88.4 m (from LM Wind Power), but by 2021 offshore turbines are expected to be 10-MW with 100 m blades. Longer blades need to be stiffer to avoid deflection, which requires materials with higher stiffness-to-weight ratio. Because the blades need to function over a 100 million load cycles over a period of 20–25 years, the fatigue of the blade materials is also critical.

## Blade Materials

Materials commonly used in wind turbine blades are described below.

### Glass and Carbon Fibers

The stiffness of composites is determined by the stiffness of fibers and their volume content. Typically, E-glass fibers are used as main reinforcement in the composites. Typically, the glass/epoxy composites for wind turbine blades contain up to 75% glass by weight. This increases the stiffness, tensile and compression strength. A promising composite material is glass fiber with modified compositions like S-glass, R-glass etc. Other glass fibers developed by Owens Corning are ECRGLAS, Advantex and WindStrand.

Carbon fiber has more tensile strength, higher stiffness and lower density than glass fiber. An ideal candidate for these properties is the spar cap, a structural element of a blade which experiences high tensile loading. A 100-m glass fiber blade could weigh up to 50 metric tons, while using carbon fiber in the spar saves 20% to 30% weight, about 15 metric tons. However, because carbon fiber is ten times more expensive, glass fiber is still dominant.

### Hybrid Reinforcements

Instead of making wind turbine blade reinforcements from pure glass or pure carbon, hybrid designs trade weight for cost. For example, for an 8 m blade, a full replacement by carbon fiber would save 80% of weight but increase costs by 150%, while a 30% replacement would save 50% of weight and increase costs by 90%. Hybrid reinforcement materials include E-glass/carbon, E-glass/aramid. The current longest blade by LM Wind Power is made of carbon/glass hybrid composites. More research is needed about the optimal composition of materials.

### Nano-engineered Polymers and Composites

Additions of small amount (0.5 weight %) of nanoreinforcement (carbon nanotubes or nanoclay) in the polymer matrix of composites, fiber sizing or interlaminar layers can improve fatigue resistance, shear or compressive strength, and fracture toughness of the composites by 30% to 80%. Research has also shown that incorporating small amounts of carbon nanotubes (CNT) can increase the lifetime up to 1500%.

### Costs

As of 2019, a wind turbine may cost around $1 million per megawatt.

For the wind turbine blades, while the material cost is much higher for hybrid glass/carbon fiber blades than all-glass fiber blades, labor costs can be lower. Using carbon fiber allows simpler designs that use less raw material. The chief manufacturing process in blade fabrication is the layering of plies. Thinner blades allow reducing the number of layers and so the labor, and in some cases, equate to the cost of labor for glass fiber blades.

## Non-blade Materials

Wind turbine parts other than the rotor blades (including the rotor hub, gearbox, frame, and tower) are largely made of steel. Smaller turbines (as well as megawatt-scale Enercon turbines) have begun using aluminum alloys for these components to make turbines lighter and more efficient. This trend may grow if fatigue and strength properties can be improved. Pre-stressed concrete has been increasingly used for the material of the tower, but still requires much reinforcing steel to meet the strength requirement of the turbine. Additionally, step-up gearboxes are being increasingly replaced with variable speed generators, which requires magnetic materials. In particular, this would require an greater supply of the rare earth metal neodymium.

Modern turbines use a couple of tons of copper for generators, cables and such. As of 2018, global production of wind turbines use 450,000 tonnes of copper per year.

## Material Supply

A study of the material consumption trends and requirements for wind energy in Europe found that bigger turbines have a higher consumption of precious metals but lower material input per kW generated. The current material consumption and stock was compared to input materials for various onshore system sizes. In all EU countries the estimates for 2020 doubled the values consumed in 2009. These countries would need to expand their resources to meet the estimated demand for 2020. For example, currently the EU has 3% of world supply of fluorspar and it requires 14% by 2020. Globally, the main exporting countries are South Africa, Mexico and China. This is similar with other critical and valuable materials required for energy systems such as magnesium, silver and indium. The levels of recycling of these materials are very low and focusing on that could alleviate supply. Because most of these valuable materials are also used in other emerging technologies, like light emitting diodes (LEDs), photo voltaics (PVs) and liquid crystal displays (LCDs), their demand is expected to grow.

A study by the United States Geological Survey estimated resources required to fulfill the US commitment to supplying 20% of its electricity from wind power by 2030. It did not consider requirements for small turbines or offshore turbines because those were not common in 2008 when the study was done. Common materials such as cast iron, steel and concrete would increase by 2%–3% compared to 2008. Between 110,000 and 115,000 metric tons of fiber glass would be required per year, a 14% increase. Rare metal use would not increase much compared to available supply, however rare metals that are also used for other technologies such as batteries which are increasing its global demand need to be taken into account. Land required would be 50,000 square kilometers onshore and 11,000 offshore. This would not be a problem in the US due to its vast area and because the same land can be used for farming. A greater challenge would be the variability and transmission to areas of high demand.

Nordex wind turbine manufacturing plant.

Permanent magnets for wind turbine generators contain rare metals such as neodymium (Nd), praseodymium (Pr), Terbium (Tb) and dysprosium (Dy). Systems that use magnetic direct drive turbines require greater amounts of rare metals. Therefore, an increase in wind turbine manufacture would increase the demand for these resources. By 2035, the demand for Nd is estimated to increase by 4,000 to 18,000 tons and for Dy by 200 to 1200 tons. These values are a quarter to half of current production. However, these estimates are very uncertain because technologies are developing rapidly.

Reliance on rare earth minerals for components has risked expense and price volatility as China has been main producer of rare earth minerals (96% in 2009) and was reducing its export quotas. However, in recent years other producers have increased production and China has increased export quotas, leading to a higher supply and lower cost, and a greater viability of large scale use of variable-speed generators.

Glass fiber is the most common material for reinforcement. Its demand has grown due to growth in construction, transportation and wind turbines. Its global market might reach US$17.4 billion by 2024, compared to US$8.5 billion in 2014. In 2014, Asia Pacific produced more than 45% of the market; now China is the largest producer. The industry receives subsidies from the Chinese government allowing it to export cheaper to the US and Europe. However, price wars have led to anti-dumping measures such as tariffs on Chinese glass fiber.

## Recycling

Interest in recycling blades varies in different markets and depends on the waste legislation and local economics. A challenge in recycling blades is related to the composite material, which is made of a thermosetting matrix and glass fibers or a combination of glass and carbon fibers. Thermosetting matrix cannot be remolded to form new composites. So the options are either to send the blade to landfill, to reuse the blade and the composite material elements found in the blade, or to transform the composite material into a new source of material. In Germany, wind turbine blades are commercially recycled as part of an alternative fuel mix for a cement factory. In the USA the town of Casper, Wyoming has buried 1,000 non-recyclable blades in its landfill site, earning $675,000 for the town. It pointed out that wind farm waste is less toxic than other garbage. Wind turbine blades represent a "vanishingly small fraction" of overall waste in the US, according to the American Wind Energy Association.

## Wind Turbines on Public Display

A few localities have exploited the attention-getting nature of wind turbines by placing them on public display, either with visitor centers around their bases, or with viewing areas farther away. The wind turbines are generally of conventional Horizontal axis, three-bladed design, and generate power to feed electrical grids, but they also serve the unconventional roles of technology demonstration, public relations, and education.

The Nordex N50 wind turbine and visitor centre.

## Small Wind Turbines

Small wind turbines may be used for a variety of applications including on- or off-grid residences, telecom towers, offshore platforms, rural schools and clinics, remote monitoring and other purposes that require energy where there is no electric grid, or where the grid is unstable. Small wind turbines may be as small as a fifty-watt generator for boat or caravan use. Hybrid solar and wind powered units are increasingly being used for traffic signage, particularly in rural locations, as they avoid the need to lay long cables from the nearest mains connection point. The U.S. Department of Energy's National Renewable Energy Laboratory (NREL) defines small wind turbines as those smaller than or equal to 100 kilowatts. Small units often have direct drive generators, direct current output, aeroelastic blades, lifetime bearings and use a vane to point into the wind.

A small Quietrevolution QR5 Gorlov type vertical axis wind turbine.
Measuring 3 m in diameter and 5 m high, it has a nameplate rating of 6.5 kW to the grid.

Larger, more costly turbines generally have geared power trains, alternating current output, and flaps, and are actively pointed into the wind. Direct drive generators and aeroelastic blades for large wind turbines are being researched.

## Wind Turbine Spacing

On most horizontal wind turbine farms, a spacing of about 6–10 times the rotor diameter is often upheld. However, for large wind farms distances of about 15 rotor diameters should be more economical, taking into account typical wind turbine and land costs. This conclusion has been reached by research conducted by Charles Meneveau of Johns Hopkins University and Johan Meyers of Leuven University in Belgium, based on computer simulations that take into account the detailed interactions among wind turbines (wakes) as well as with the entire turbulent atmospheric boundary layer.

Recent research by John Dabiri of Caltech suggests that vertical wind turbines may be placed much more closely together so long as an alternating pattern of rotation is created allowing blades of neighbouring turbines to move in the same direction as they approach one another.

## Operability

### Maintenance

Wind turbines need regular maintenance to stay reliable and available. In the best case turbines are available to generate energy 98% of the time.

Modern turbines usually have a small onboard crane for hoisting maintenance tools and minor components. However, large, heavy components like generator, gearbox, blades, and so on are rarely replaced, and a heavy lift external crane is needed in those cases. If the turbine has a difficult access road, a containerized crane can be lifted up by the internal crane to provide heavier lifting.

## Repowering

Installation of new wind turbines can be controversial. An alternative is repowering, where existing wind turbines are replaced with bigger, more powerful ones, sometimes in smaller numbers while keeping or increasing capacity.

## Demolition

Older turbines were in some early cases not required to be removed when reaching the end of their life. Some still stand, waiting to be recycled or repowered.

A demolition industry develops to recycle offshore turbines at a cost of DKK 2–4 million per (MW), to be guaranteed by the owner.

## Comparison with Fossil-fuel Turbines

### Advantages

Wind turbines produce electricity at between two and six cents per kilowatt hour, which is one of the lowest-priced renewable energy sources. As technology needed for wind turbines continued to improve, the prices decreased as well. In addition, there is currently no competitive market for wind energy, because wind is a freely available natural resource, most of which is untapped. The main cost of small wind turbines is the purchase and installation process, which averages between $48,000 and $65,000 per installation. The energy harvested from the turbine will offset the installation cost, as well as provide virtually free energy for years.

Wind turbines provide a clean energy source, use little water, emitting no greenhouse gases and no waste products. Over 1,500 tons of carbon dioxide per year can be eliminated by using a one-megawatt turbine instead of one megawatt of energy from a fossil fuel.

### Disadvantages

Wind turbines can be very large, reaching over 140 m (460 ft) tall and with blades 55 m (180 ft) long, and people have often complained about their visual impact.

Environmental impact of wind power includes effect on wildlife, but can be mitigated if proper monitoring and mitigation strategies are implemented. Thousands of birds, including rare species, have been killed by the blades of wind turbines, though wind turbines contribute relatively insignificantly to anthropogenic avian mortality. Wind farms and nuclear power stations are responsible for between 0.3 and 0.4 bird deaths per gigawatt-hour (GWh) of electricity while fossil fueled power stations are responsible for about

5.2 fatalities per GWh. In 2009, for every bird killed by a wind turbine in the US, nearly 500,000 were killed by cats and another 500,000 by buildings. In comparison, conventional coal fired generators contribute significantly more to bird mortality, by incineration when caught in updrafts of smoke stacks and by poisoning with emissions byproducts (including particulates and heavy metals downwind of flue gases). Further, marine life is affected by water intakes of steam turbine cooling towers (heat exchangers) for nuclear and fossil fuel generators, by coal dust deposits in marine ecosystems (e.g. damaging Australia's Great Barrier Reef) and by water acidification from combustion monoxides.

Energy harnessed by wind turbines is intermittent, and is not a "dispatchable" source of power; its availability is based on whether the wind is blowing, not whether electricity is needed. Turbines can be placed on ridges or bluffs to maximize the access of wind they have, but this also limits the locations where they can be placed. In this way, wind energy is not a particularly reliable source of energy. However, it can form part of the energy mix, which also includes power from other sources. Notably, the relative available output from wind and solar sources is often inversely proportional (balancing). Technology is also being developed to store excess energy, which can then make up for any deficits in supplies.

## Airborne Wind Turbine

An airborne wind turbine is a design concept for a wind turbine with a rotor supported in the air without a tower, thus benefiting from more mechanical and aerodynamic options, the higher velocity and persistence of wind at high altitudes, while avoiding the expense of tower construction, or the need for slip rings or yaw mechanism. An electrical generator may be on the ground or airborne. Challenges include safely suspending and maintaining turbines hundreds of meters off the ground in high winds and storms, transferring the harvested and generated power back to earth, and interference with aviation.

Kiwee one: An Airborne wind turbine.

Airborne wind turbines may operate in low or high altitudes; they are part of a wider class of Airborne Wind Energy Systems (AWES) addressed by high-altitude wind power and crosswind kite power. When the generator is on the ground, then the tethered aircraft need not carry the generator mass or have a conductive tether. When the generator is aloft, then a conductive tether would be used to transmit energy to the ground or used aloft or beamed to receivers using microwave or laser. Kites and helicopters come down when there is insufficient wind; kytoons and blimps may resolve the matter with other disadvantages. Also, bad weather such as lightning or thunderstorms, could temporarily suspend use of the machines, probably requiring them to be brought back down to the ground and covered. Some schemes require a long power cable and, if the turbine is high enough, a prohibited airspace zone. As of July 2015, no commercial airborne wind turbines are in regular operation.

## Aerodynamic Variety

An aerodynamic airborne wind power system relies on the wind for support.

Miles L. Loyd proposed and analyzed an efficient AWES in his work "Crosswind Kite Power" in 1980. Power output of AWES with crosswind wing motion is proportional to a square of a lift/drag ratio of the wing. Such AWES is based on the same aerodynamic principles as a conventional wind turbine (AWES), but it is more efficient because the air speed is constant along the wing span and the aerodynamic forces are resisted by tension of a tether, rather than by bending of a tower.

Crosswind kite generator with fast motion transfer.

Bryan Roberts, a professor of engineering at the University of Technology, in Sydney, Australia, has proposed a helicopter-like craft which flies to 15,000 feet (4,600 m) altitude and stays there, held aloft by wings that generate lift from the wind, and held in place by a cable to a ground anchor. According to its designers, while some of the energy in the wind would be 'lost' on lift, the constant and potent winds would allow it to generate constant electricity. Since the winds usually blow horizontally, the turbines would be at an angle from the horizontal, catching winds while still generating lift.

Deployment could be done by feeding electricity to the turbines, which would turn them into electric motors, lifting the structure into the sky.

The Dutch ex-astronaut and physicist Wubbo Ockels, working with the Delft University of Technology in the Netherlands, has designed and demonstrated an airborne wind turbine he called a "Laddermill". It consists of an endless loop of kites. The kites lift up one end of the endless loop (the "ladder"), and the released energy is used to drive an electric generator.

A September 2009 paper from Carbon Tracking Ltd., Ireland has shown the capacity factor of a kite using ground-based generation to be 52.2%, which is better than terrestrial wind-farm capacity factors of 30%.

A team from Worcester Polytechnic Institute in the United States has developed a smaller-scale with an estimated output of about 1 kW. It uses a kiteboarding kite to induce a rocking motion in a pivoting beam.

The KiteGen uses a prototype vertical-axis wind turbine (actually a semi-rigid cross wind kite). It is an innovative plan (still in the construction phase) that consists of one wind farm with a vertical spin axis, and employs kites to exploit high-altitude winds. The Kite Wind Generator (KWG) or KiteGen is claimed to eliminate all the static and dynamic problems that prevent the increase of the power (in terms of dimensions) obtainable from the traditional Horizontal axis wind turbine generators. Generating equipment would remain on the ground, and only the airfoils are supported by the wind. Such a wind power plant would be capable of producing the energy equivalent to a nuclear plant, while using an area of few square kilometres, without occupying it exclusively. (The majority of this area can still be used for agriculture, or navigation in the case of an offshore installation.)

The Rotokite is developed from Gianni Vergnano's idea. It uses aerodynamic profiles similar to kites that have been rotated on their own axis, emulating the performance of a propeller. The use of the rotation principle simplifies the problem of checking the flight of the kites and eliminates the difficulties due to the lengths of cables, enabling the production of wind energy at low cost. The Heli Wind Power is a project of Gianni Vergnano that uses a tethered kite.

The HAWE System is developed from Tiago Pardal's idea. The system that consists in a pumping cycle similar to kite systems. In Generation Phase the pulling force increase 5-10 times due to Magnus Effect of a spinning cylinder(aerial platform), like a kite the pulling force produced by the aerial platform will unwind the cable and generate electricity in the ground. In the recovery phase it rewinds the cable with no Magnus effect in the aerial platform.

In August 2011 the German company SkySails, producer of kites for ship propulsion, announced a kite-based wind power system for on- and offshore applications that is supposed to be "30% cheaper than current offshore solutions".

In June 2012, the German company NTS GmbH had successfully tested X-Wind technology (spoken: Cross-Wind) on linear rail system in Freidland, Germany. "NTS Energie- und Transportsysteme GmbH" was found in 2006 by Uwe Ahrens. X-Wind technology combines two well-known technologies - automatically steered kites and generators on a closed loop rail system. Closed loop prototype is under construction at Mecklenburg-Vorpommern, Germany. This technology allows to harness increasingly stable and constant wind currents at altitudes between 200 and 500m. Technical report readings and measurements show that NTS X-Wind Systems double to triple the efficiency of conventional wind energy systems according to energy production.

In May 2013, the Californian company Makani Power, developer of some crosswind hybrid kite systems with onboard generator doubling as motor, has been acquired by Google.

In May 2013, an airborne wind energy system with a ground-based generator using fast motion transfer was suggested by L. Goldstein.

In 2015, a sails on rope wind and ocean current energy system was invented by a Taiwanese Johnson Hsu.

On 15 December 2015 Windswept and Interesting Ltd demonstrated a "Daisy" kite ring stack airborne wind turbine. The Daisy kite stack demonstrated on 15 December 2015 is the only airborne wind energy system to have won the someawe.org 100*3 AWE challenge. The Daisy system uses tensioned torsion transfer of kite motion to turn a ground based generator.

Kitemill flies an airplane tethered to a winch powering a 5 kW generator on ground.

Since 2014, Kitewinder, a French company located near Bordeaux is working on the first commercial airborne wind turbine named Kiwee One. Kiwee one is a back pack airborne wind turbine for nomadic uses with a nominal power of 100 Watts. Kiwee one launch procedure is manual but the product is equipped with an automatic retrieval mechanism for low wind/ high wind conditions.

## Aerostat Variety

An aerostat-type wind power system relies at least in part on buoyancy to support the wind-collecting elements. Aerostats vary in their designs and resulting lift-to-drag ratio; the kiting effect of higher lift-over-drag shapes for the aerostat can effectively keep an airborne turbine aloft; a variety of such kiting balloons were made famous in the kytoon by Domina Jalbert.

Balloons can be incorporated to keep systems up without wind, but balloons leak slowly and have to be resupplied with lifting gas, possibly patched as well. Very large, sun-heated balloons may solve the helium or hydrogen leakage problems.

An Ontario based company called Magenn is developing a turbine called the Magenn Air Rotor System (MARS). A future 1,000-foot (300 m)-wide MARS system would use a horizontal rotor in a helium suspended apparatus which is tethered to a transformer on the ground. Magenn claims that their technology provides high torque, low starting speeds, and superior overall efficiency thanks to its ability to deploy higher in comparison to non-aerial solutions. The first prototypes were built by TCOM in April 2008. No production units have been delivered.

Boston-based Altaeros Energies uses a helium-filled balloon shroud to lift a wind turbine into the air, transferring the resultant power down to a base station through the same cables used to control the shroud. A 35-foot prototype using a standard Skystream 2.5kW 3.7m wind turbine was flown and tested in 2012. In fall 2013, Altaeros was at work on its first commercial-scale demonstration in Alaska.

## Floating Wind Turbine

A floating wind turbine is an offshore wind turbine mounted on a floating structure that allows the turbine to generate electricity in water depths where fixed-foundation turbines are not feasible. Floating wind farms have the potential to significantly increase the sea area available for offshore wind farms, especially in countries with limited shallow waters, such as Japan. Locating wind farms further offshore can also reduce visual pollution, provide better accommodation for fishing and shipping lanes, and reach stronger and more consistent winds.

The world's first full-scale floating wind turbine, Hywind.

Commercial floating wind turbines are mostly at the early phase of development, with several single turbine prototypes having been installed since 2007. As of 2018, the only operational floating wind farm is Hywind Scotland, developed by Equinor ASA and commissioned in October 2017. The farm has 5 floating turbines with a total capacity of 30 MW.

The world's second full-scale floating wind turbine (and first to be installed without the use of heavy-lift vessels), WindFloat, operating at rated capacity (2 MW) approximately 5 km offshore.

University of Maine's VolturnUS 1:8 was the first grid-connected offshore wind turbine.

The concept for large-scale offshore floating wind turbines was introduced by Professor William E. Heronemus at the University of Massachusetts Amherst in 1972. It was not until the mid 1990s, after the commercial wind industry was well established, that the topic was taken up again by the mainstream research community.

Blue H Technologies of the Netherlands deployed the first 80-kW floating wind turbine 21.3 kilometres (13.2 mi) off the coast of Apulia, Italy in December 2007. The prototype was installed in waters 113 m deep in order to gather test data on wind and sea conditions, and was decommissioned at the end of 2008. The turbine utilized a tension-leg platform design and a two-bladed turbine.

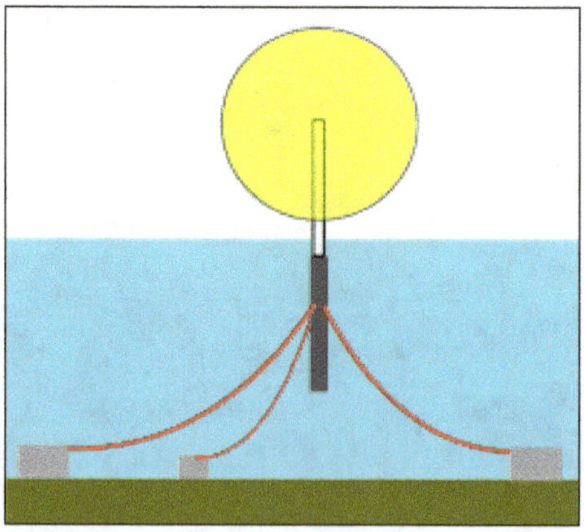

A single floating cylindrical spar buoy moored by catenary cables. Hywind uses a ballasted catenary layout that adds 60 tonne weights hanging from the midpoint of each anchorcable to provide additional tension.

The first large-capacity, 2.3-megawatt floating wind turbine was Hywind, which became operational in the North Sea near Norway in September 2009. The turbine was constructed by Siemens Wind Power and mounted on a floating tower with a 100 m deep draft, with a float tower constructed by Technip.

After assembly in the calmer waters of Åmøy Fjord near Stavanger, Norway, the 120 m tall tower was towed 10 km offshore into 220 m deep water, 10 km southwest of Karmøy, on 6 June 2009 for a two-year test deployment. Alexandra Beck Gjorv of Statoil said, "[The experiment] should help move offshore wind farms out of sight. The global market for such turbines is potentially enormous, depending on how low we can press costs." Hywind, owned by Statoil, cost 400 million kroner (around US$62 million) to build and deploy. The 13-kilometre (8.1 mi) long submarine power transmission cable was installed in July 2009 and system test including rotor blades and initial power transmission was conducted shortly thereafter. The installation was expected to generate about 9 GW·h of electricity annually. In 2010 it survived 11 meter waves with seemingly no wear. By 2016, the turbine had produced 50 GWh; an overall capacity factor of 41%.

In October 2011, a WindFloat prototype was installed 4 km offshore of Aguçadoura, Portugal by Principle Power in approximately 45 m of water (previously the Aguçadoura Wave Farm site). The WindFloat was fitted with a Vestas V80 2.0-MW offshore wind turbine and grid connected. The installation was the first offshore wind turbine to be deployed without the use of any offshore heavy lift vessels as the turbine was fully commissioned onshore prior to the unit's being towed offshore. This is the first offshore wind turbine installed in open Atlantic waters, and the first to employ a semi-submersible type floating foundation.

By January 2020, the first of WindFloat's three 8.4-megawatt MHI Vestas turbines was in operation. Power is transmitted to a substation 12 miles away on shore, through a cable anchored to the seabed at a depth of about 100 meters.

SeaTwirl deployed its first floating grid connected wind turbine off the coast of Sweden in August 2011. It was tested and decommissioned. The design intended to store energy in a flywheel, thus, energy could be produced even after the wind stopped blowing.

In June 2013, the University of Maine deployed the VolturnUS 1:8, a 65 foot (20 m) tall floating turbine prototype that is 1:8th the scale of a 6-megawatt (MW), 450 foot (140 m) rotor diameter design. VolturnUS 1:8 was the first grid-connected offshore wind turbine deployed in the Americas. The VolturnUS design utilizes a concrete semi-submersible floating hull and a composite materials tower designed to reduce both capital and Operation and Maintenance costs, and to allow local manufacturing. The technology was the result of collaborative research and development conducted by the University of Maine-led DeepCwind Consortium.

In 2013, Statoil pulled out of the $120 million project of four 3-MW turbines floating in 140 m depth of water near Boothbay Harbor, Maine citing change in legislation, and focused on their five 6-MW turbines in Scotland instead, where the average wind speed is 10 m/s and the water depth is 100 m.

In 2015, Statoil received permission to install Hywind Scotland, a 30 MW floating wind farm 18 miles (29 km) off Peterhead using 5 Hywind turbines. Statoil also plans to test a 1 MWh lithium-ion battery system (called Batwind) with the farm. Hywind Scotland was commissioned in October 2017, becoming the first commercial floating wind farm in operation. Construction of the NOK 2 billion (£152m) project started in 2016 in Spain, Norway and Scotland. The turbines were assembled at Stord in Norway in summer 2017 using the Saipem 7000 floating crane, and the finished turbines were moved to near Peterhead. Three suction cup anchors hold each turbine.

The first floating turbine in Japan was floated near Fukue Island in 2016, after a 5-year demonstration period near shore. The 2-MW turbine was developed by Hitachi.

In June 2016, Maine's New England Aqua Ventus I floating offshore wind demonstration project, designed by the DeepCwind Consortium, was selected by the U.S. Department of Energy to participate in the Offshore Wind Advanced Technology Demonstration program.

## Anchoring Systems

Two common types of engineered design for anchoring floating structures include tension-leg and catenary loose mooring systems. Tension leg mooring systems have vertical tethers under tension providing large restoring moments in pitch and roll. Catenary mooring systems provide station–keeping for an offshore structure yet provide little stiffness at low tensions." A third form of mooring system is the ballasted catenary configuration, created by adding multiple-tonne weights hanging from the midsection of each anchor cable in order to provide additional cable tension and therefore increase stiffness of the above-water floating structure.

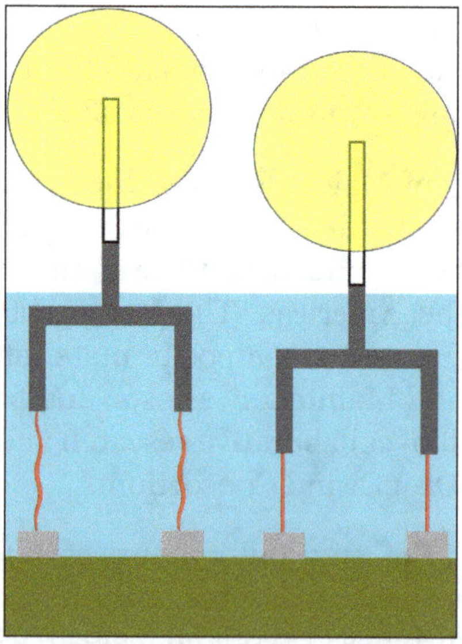

Left-hand tower-bearing structure (grey) is free floating, the right-hand structure is pulled by the tensioned cables (red) down towards the seabed anchors (light-grey).

The IEC 61400–3 design standard requires that a loads analysis be based on site-specific external conditions such as wind, wave and currents. The IEC 61400–3-2 standard applies specifically to floating wind turbines.

## Economics

The technical feasibility of deepwater floating wind turbines is not questioned, as the long-term survivability of floating structures has been successfully demonstrated by the marine and offshore oil industries over many decades. However, the economics that allowed the deployment of thousands of offshore oil rigs have yet to be demonstrated for floating wind turbine platforms. For deepwater wind turbines, a floating structure will replace pile-driven monopoles or conventional concrete bases that are commonly used as foundations for shallow water and land-based turbines. The floating structure must provide enough buoyancy to support the weight of the turbine and to restrain pitch, roll and heave motions within acceptable limits. The capital costs for the wind turbine itself will not be significantly higher than current marine-proofed turbine costs in shallow water. Therefore, the economics of deepwater wind turbines will be determined primarily by the additional costs of the floating structure and power distribution system, which are offset by higher offshore winds and close proximity to large load centres (e.g. shorter transmission runs).

As of 2009 however, the economic feasibility of shallow-water offshore wind technologies is more completely understood. With empirical data obtained from fixed-bottom installations off many countries since the late 1990s, representative costs are well understood. Shallow-water turbines cost US$2.4-3 million per megawatt to install, according to the World Energy Council, while the practical feasibility and per-unit economics of deep-water, floating-turbine offshore wind is yet to be established. Initial deployment of single full-capacity turbines in deep-water locations began only in 2009.

As of October 2010, new feasibility studies are supporting that floating turbines are becoming both technically and economically viable in the UK and global energy markets. "The higher up-front costs associated with developing floating wind turbines would be offset by the fact that they would be able to access areas of deep water off the coastline of the UK where winds are stronger and reliable." The recent Offshore Valuation study conducted in the UK has confirmed that using just one third of the UK's wind, wave and tidal resource could generate energy equivalent to 1 billion barrels of oil per year; the same as North Sea oil and gas production. A significant challenge when using this approach is the coordination needed to develop transmission lines.

A 2015 report by Carbon Trust recommends 11 ways to reduce cost. Also in 2015, researchers at University of Stuttgart estimated cost at €230/MWh.

## Oil Well Injection

When oil fields become depleted, the operator injects water to keep pressure high for secondary recovery. This requires power, but installing gas turbines means shutting down the extraction process, losing valuable income. The classification society DNV GL has calculated that in some cases a floating wind turbine can economically provide power for injection, as the oil platform can keep on producing, avoiding a costly pause.

In 2016 DNV GL, ExxonMobil and others approved calculations of saving $3/barrel of oil using a 6MW Hywind instead of traditional engines, driving two 2MW pumps injecting water into an offshore oil well. At least 44,000 barrels of processed water per day can be injected, even on calm June days. The project began laboratory testing in 2017.

## Floating Wind Farms

Hywind Scotland has 5 floating turbines with a total capacity of 30 MW, and operated since 2017. Japan has 4 floating turbines with a combined 16 MW capacity.

## Proposals

In 2011, Japan planned to build a pilot floating wind farm, with six 2-megawatt turbines, off the Fukushima coast of northeast Japan where the Fukushima Daiichi nuclear disaster created a scarcity of electric power. After the evaluation phase is complete in 2016, "Japan plans to build as many as 80 floating wind turbines off Fukushima by 2020." The cost is expected to be in the range of 10–20 billion Yen over five years to build the first six floating wind turbines. In 2011, some foreign companies had also planned to bid on the 1-GW large floating wind farm that Japan hoped to build by 2020. In March 2012, Japan's Ministry of Economy, Trade and Industry approved a 12.5bn yen ($154m) project to float a 2-MW Fuji in March 2013 and two 7-MW Mitsubishi hydraulic "SeaAngel" later about 20–40 km offshore in 100–150 metres of water depth. The Japanese Wind Power Association claims a potential of 519 GW of floating offshore wind capacity in Japan. The first turbine became operational in November 2013. In 2018, NEDO has announced two tenders to be launched in early February, aiming to support the development of both floating and fixed-bottom offshore wind projects in the country.

The US State of Maine solicited proposals in September 2010 to build a floating wind farm. The RFPis seeking proposals for 25 MW of deep-water offshore wind capacity to supply power for 20-year long-term contract period in the Gulf of Maine. Proposals were due by May 2011. In April 2012 Statoil received state regulatory approval to build a large four-unit demonstration wind farm off the coast of Maine. As of April 2013, the Hywind 2 4-tower, 12–15 MW wind farm was being developed by Statoil North America for placement 20 kilometres (12 mi) off the east coast of Maine in 140–158 metres (459–518 ft)-deep water of the Atlantic Ocean. Like the first Hywind installation off Norway, the turbine foundation will be a spar floater. The State of Maine Public Utility

Commission voted to approve the construction and fund the US$120 million project by adding approximately 75 cents/month to the average retail electricity consumer. Power could be flowing into the grid no earlier than 2016. As a result of legislation in 2013 by the State of Maine, Statoil placed the planned Hywind Maine floating wind turbine development project on hold in July 2013. The legislation required the Maine Public Utilities Commission to undertake a second round of bidding for the offshore wind sites with a different set of ground rules, which subsequently led Statoil to suspend due to increased uncertainty and risk in the project. Statoil is considering other locations for its initial US demonstration project. Some vendors who could bid on the proposed project in Maine expressed concerns in 2010 about dealing with the United States regulatory environment. Since the proposed site is in federal waters, developers would need a permit from the US Minerals Management Service, "which took more than seven years to approve a yet-to-be-built, shallow-water wind project off Cape Cod" (Cape Wind). "Uncertainty over regulatory hurdles in the United States is 'the Achilles heel' for Maine's ambitions for deepwater wind."

In August 2019, Enova awarded NOK2.3 billion to Equinor for a NOK5 billion 88 MW floating wind farm called Hywind Tampen, with the purpose of reducing technology costs and supplying power to the Snorre and Gullfaks oil fields from 2022.

## Other Applications

As they are suitable for towing, floating wind turbine units can be relocated to any location on the sea without much additional cost. So they can be used as prototype test units to practically assess the design adequacy and wind power potential of prospective sites.

When the transmission of generated wind power to nearby land is not economical, the power can be used in power to gas applications to produce hydrogen gas, ammonia/ urea, reverse osmosis water desalination, natural gas, LPG, alkylate/ gasoline, etc. on floating platforms which can be easily transported to nearby consuming centres.

Floating wind turbines can be used to provide motive power for achieving artificial upwelling of nutrient-rich deep ocean water to the surface for enhancing fisheries growth in areas with tropical and temperate weather. Though deep seawater (below 50 metres depth) is rich in nutrients such as nitrogen and phosphorus, the phytoplankton growth is poor due to the absence of sunlight. The most productive ocean fishing grounds are located in cold water seas at high latitudes where natural upwelling of deep sea water occurs due to inverse thermocline temperatures. The electricity generated by the floating wind turbine would be used to drive high–flow and low–head water pumps to draw cold water from below 50 meters water depth and mixed with warm surface water by eductors before releasing into the sea. Mediterranean Sea, Black Sea, Caspian Sea, Red Sea, Persian Gulf, deep water lakes/reservoirs are suitable for artificial upwelling for enhancing fish catch economically. These units can also be mobile type to utilise the seasonal favourable winds all around the year.

## Floating Design Concepts

### Deep Wind

Risø DTU National Laboratory for Sustainable Energy and 11 international partners started a 4-year program called DeepWind in October 2010 to create and test economical floating Vertical Axis Wind Turbines up to 20 MW. The program is supported with €3 million through EUs Seventh Framework Programme. Partners include TUDelft, Aalborg University, SINTEF, Equinor and United States National Renewable Energy Laboratory.

### Flowocean

Flowocean is a Swedish technology company with its own proprietary technology for floating offshore wind power with head office in the city of Västerås, Sweden. FLOW is a semi-submersible floating offshore wind turbine technology with two wind turbine generators on one floating platform. The structure weather vanes passively so that the wind turbines always face the wind. Flow technology is a combination of Tension Leg Platform(TLP) and Semi-Submersible which gives the Flow unit the benefits of both principles and allows the unit to be robust and light.

Flowocean has developed a patented design for floating offshore wind power plants aiming to make floating offshore wind power cost-effective. FLOW can be considered an assembly of three systems, the floater, the buoy and the mooring system. The floater is all structure that is rotating. The buoy is of turret type, is moored to the sea bed and contains a bearing that allows the floater to rotate freely around it. The mooring system is the set of components that anchors the buoy to the sea bed, i.e. mooring lines/ropes/chains, chain stoppers and anchors. The FLOW units are highly standardised with all sub-systems well proven. Inter-array wind farm cabling and mooring systems are shared between the units.

### GICON

The GICON-TLP is a floating substructure system based on a tension leg platform (TLP) developed by GICON GmbH. The system is deployable from 45 meters to 350 meters water depth. It consists of six major components: four buoyancy bodies, horizontal pipes for structural base, vertical pipes which pass through the water line, angled piles for connection with the transition piece. Cast nodes are used to connect all components. The TLP can be equipped with an offshore wind turbine in range of 6–10 MW.

The GICON-TLP is anchored to the seabed via four pre-tensioned mooring ropes with a buoyant gravity-base anchor consisting of concrete. No pile driving or drilling for anchoring is necessary. All ropes are connected at the corners of the square based system. The TLP for a 6MW wind turbine is currently being developed by the GICON Group and their key partner, the Endowed Chair for Wind Energy Technology (LWET) at the University of Rostock, utilizing prefabricated steel-concrete composite components in combination with components of steel. A main focus of the TLP design is on the modularity and the possibility of assembly in any dry dock near to the installation site and without the use of construction vessels. After offshore location is reached, joints of TLP and anchor will be decoupled and the gravity anchor will be lowered down by using ballast water. Once the anchor has reached the bottom, it is filled with sand. One unique feature of the system is the sufficient floating stability during transport as well as during operations.

In October 2017, model tests took place in the model test facility of French École Centrale de Nantes (ECN) with 1:50 model of the GICON-TLP incl. wind turbine. Based on this test a TRL of 5 was reached.

## Ideol

Ideol's engineers have developed and patented a ring-shaped floating foundation based on a central opening system (Damping Pool) used for optimizing foundation + wind turbine stability. As such, the sloshing water contained in this central opening counteracts the swell-induced floater oscillations. Foundation-fastened mooring lines are simply attached to the seabed to hold the assembly in position. This floating foundation is compatible with all wind turbines without any modification and has reduced dimensions (from 36 to 55 metres per side for a wind turbine between 2 and 8 MW). Manufacturable in concrete or steel, this floating foundation allows for local construction near project sites.

Steel floating substructure designed by Ideol for 3.2 MW NEDO project based on Ideol technology, fully coated in dry dock before wind turbine installation.

Ideol's 2MW floating wind turbine.

Ideol leads the FLOATGEN project, a floating wind turbine demonstration project based on Ideol's technology, built by Bouygues Travaux Publics and operational off the coast of Le Croisic on the offshore experimentation site of Ecole Centrale de Nantes (SEM-REV). The construction of this project, France's first offshore wind turbine with a capacity of 2 MW, was completed in April 2018 and the unit installed on site in August 2018.

In August 2018, Hibiki, the second demonstrator with an Aerodyn 3.2 MW wind turbine was installed 15 km East of the port of Kitakyushu by Japanese conglomerate

Hitachi Zosen. Ideol developed the design for this steel hull that was manufactured in a Japanese dry dock.

In August 2017, the French government has selected Eolmed, a consortium led by French renewable energy developer Quadran in association with Ideol, Bouygues Travaux Publics and Senvion, for the development and construction of a 25MW Mediterranean floating offshore wind farm 15 km off the coastal town of Gruissan (Languedoc-Roussillon), planned to be commissioned 2020.

## Nautica Windpower

Nautica Windpower has proposed a technique for potentially reducing system weight, complexity and costs for deepwater sites. Scale model tests in open water have been conducted in Lake Erie and structural dynamics modeling was done in 2010 for larger designs. Nautica Windpower's Advanced Floating Turbine (AFT) uses a single mooring line and a downwind two-bladed rotor configuration that is deflection tolerant and aligns itself with the wind without an active yaw system. Two-bladed, downwind turbine designs that can accommodate flexibility in the blades will potentially prolong blade lifetime, diminish structural system loads and reduce offshore maintenance needs, yielding lower lifecycle costs.

## SeaTwirl

SeaTwirl develops a floating vertical axis wind turbine (VAWT). The floater is based on a SPAR solution and is rotating along with the turbine. The concept limits the need for moving parts as well as bearings in the hub region. SeaTwirl is based in Gothenburg Sweden and is registered on the European growth market First North. 2015 SeaTwirl launched a 30 kW prototype in the archipelago of Sweden which is connected to the grid at Lysekil. The aim is to scale the concept and to prove the scalability with a turbine of 1MW size in 2020. The concept is scalable for sizes well over 10MW.

## VolturnUS

The VolturnUS design utilizes a concrete semi-submersible floating hull and a composite materials tower designed to reduce both capital and Operation and Maintenance costs, and to allow local manufacturing.

VolturnUS is North America's first floating grid-connected wind turbine. It was lowered into the Penobscot River in Maine on 31 May 2013 by the University of Maine Advanced Structures and Composites Center and its partners. During its deployment, it experienced numerous storm events representative of design environmental conditions prescribed by the American Bureau of Shipping (ABS) Guide for Building and Classing Floating Offshore Wind Turbines, 2013.

The VolturnUS design utilizes a concrete semi-submersible floating hull and a composite materials tower designed to reduce both capital and Operation and Maintenance costs, and to allow local manufacturing.

VolturnUS is North America's first floating grid-connected wind turbine. It was lowered into the Penobscot River in Maine on 31 May 2013 by the University of Maine Advanced Structures and Composites Center and its partners. During its deployment, it experienced numerous storm events representative of design environmental conditions prescribed by the American Bureau of Shipping (ABS) Guide for Building and Classing Floating Offshore Wind Turbines, 2013.

The VolturnUS floating concrete hull technology can support wind turbines in water depths of 45 m or more. With 12 independent cost estimates from around the U.S. and the world, it has been found to significantly reduce costs compared to existing floating systems. The design has also received a complete third-party engineering review.

In June 2016, the UMaine-led New England Aqua Ventus I project won top tier status from the US Department of Energy (DOE) Advanced Technology Demonstration Program for Offshore Wind. This means that the Aqua Ventus project is now automatically eligible for an additional $39.9 Million in construction funding from the DOE, as long as the project continues to meet its milestones.

## WindFloat

WindFloat is a floating foundation for offshore wind turbines designed and patented by Principle Power. A full-scale prototype was constructed in 2011 by Windplus, a joint-venture between EDP, Repsol, Principle Power, A. Silva Matos, Inovcapital, and FAI. The complete system was assembled and commissioned onshore including the turbine. The entire structure was then wet-towed 400 kilometres (250 mi) (from southern to northern Portugal) to its final installed location 5 kilometres (3.1 mi) offshore of Aguçadoura, Portugal, previously the Aguçadoura Wave Farm. The WindFloat was equipped with a Vestas v80 2.0-megawatt turbine and installation was completed on 22 October 2011. A year later, the turbine had produced 3 GWh. The cost of this project is around €20 million (about US$26 million). This single wind turbine can produce energy to power 1300 homes.

The subsea metal structure is reported to improve dynamic stability, whilst still maintaining shallow draft, by dampening wave– and turbine–induced motion utilizing a tri-column triangular platform with the wind turbine positioned on one of the three columns. The triangular platform is then “moored” using a conventional catenary mooring consisting of four lines, two of which are connected to the column supporting the turbine, thus creating an “asymmetric mooring.” As the wind shifts direction and changes the loads on the turbine and foundation, a secondary hull-trim system shifts ballast water between each of the three columns. This permits the platform to maintain even keel while producing the maximum amount of energy. This is in contrast to other floating concepts which have implemented control strategies that de-power the turbine to compensate for changes in turbine thrust-induced overturning moment. This technology could allow wind turbines to be sited in offshore areas that were previously considered inaccessible, areas having water depth exceeding 40 m and more powerful wind resources than shallow-water offshore wind farms typically encounter.

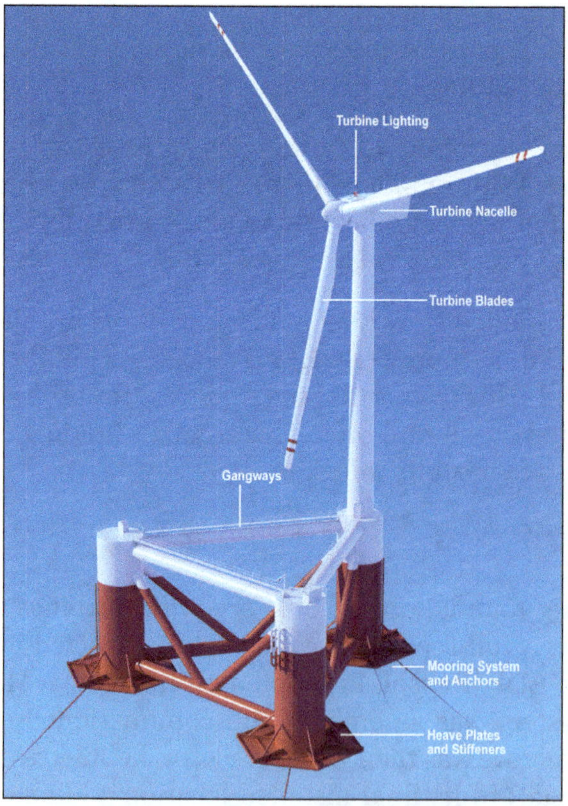


A diagram of the WindFloat system.

A 25 MW WindFloat project received government permission in December 2016, with EU funding the €48 million transmission cable. The €100 million project is expected to be funded by 2017 and operational by 2019. A WindFloat with a 2MW Vestas turbine installed near Scotland began delivering power in late 2018.

Principle Power was planning a 30-MW WindFloat project using 6-MW Siemens turbines in 366 m of water near Coos Bay, Oregon to be operational in 2017, but the project has since been cancelled.

### Others

A combined floating wave and wind power plant was installed at Vindeby Offshore Wind Farm in 2010.

An open source project was proposed by former Siemens director Henrik Stiesdal in 2015 to be assessed by DNV GL. It suggests using tension leg platforms with replaceable pressurized tanks anchored to sheet walls.

The International Energy Agency (IEA), under the auspices of their Offshore Code Comparison Collaboration (OC3) initiative, has completed high-level design and simulation modeling of the OC-3 Hywind system, a 5-MW wind turbine installed on a floating spar buoy, moored with catenary mooring lines, in water depth of 320 metres. The spar buoy platform would extend 120 metres below the surface and the mass of such a system, including ballast would exceed 7.4 million kg.

VertiWind is a floating vertical axis wind turbine design created by Nenuphar whose mooring system and floater are designed by Technip.

PivotBuoy received €4m EU funding in 2019 to install a 225kW downwind turbine at the Oceanic Platform of the Canary Islands.

## Vertical Axis Wind Turbine

A vertical-axis wind turbines (VAWT) is a type of wind turbine where the main rotor shaft is set transverse to the wind (but not necessarily vertically) while the main components are located at the base of the turbine. This arrangement allows the generator and gearbox to be located close to the ground, facilitating service and repair. VAWTs do not need to be pointed into the wind, which removes the need for wind-sensing and orientation mechanisms. Major drawbacks for the early designs (Savonius, Darrieus and giromill) included the significant torque variation or "ripple" during each revolution, and the large bending moments on the blades. Later designs addressed the torque ripple issue by sweeping the blades helically (Gorlov type).

A vertical axis wind turbine has its axis perpendicular to the wind streamlines and vertical to the ground. A more general term that includes this option is "transverse axis wind turbine" or "cross-flow wind turbine." For example, the original Darrieus patent, US Patent 1835018, includes both options.

The world's tallest vertical-axis wind turbine.

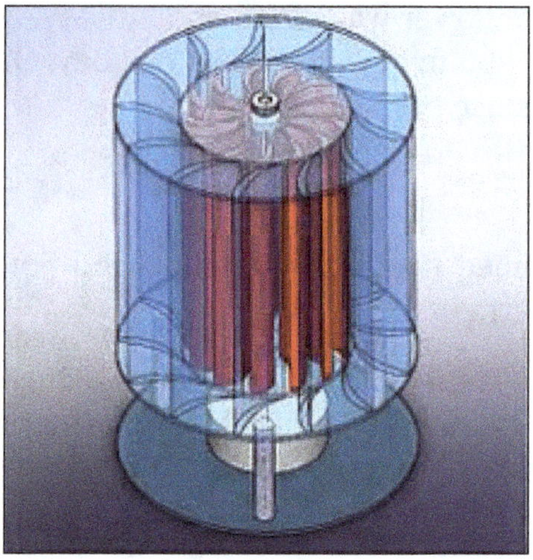

Vortexis Schematic.

Drag-type VAWTs such as the Savonius rotor typically operate at lower tipspeed ratios than lift-based VAWTs such as Darrieus rotors and cycloturbines.

## General Aerodynamics

The forces and the velocities acting in a Darrieus turbine are depicted in figure. The resultant velocity vector, W → $\vec{W}$, is the vectorial sum of the undisturbed upstream air velocity, U → $\vec{U}$, and the velocity vector of the advancing blade, − ω → × R → $-\vec{\omega}\times\vec{R}$.

$$\vec{W}=\vec{U}+(-\vec{\omega}\times\vec{R})$$

Thus the oncoming fluid velocity varies during each cycle. Maximum velocity is found for $\theta = 0°$ and the minimum is found for $\theta = 180°$, where $\theta$ is the azimuthal or orbital blade position. The angle of attack, $\alpha$, is the angle between the oncoming air speed, W, and the blade's chord. The resultant airflow creates a varying, positive angle of attack to the blade in the upstream zone of the machine, switching sign in the downstream zone of the machine.

It follows from geometric considerations of angular velocity as seen in the accompanying figure that:

$$V_t = R\omega + U\cos(\theta)$$

and

$$V_n = U\sin(\theta)$$

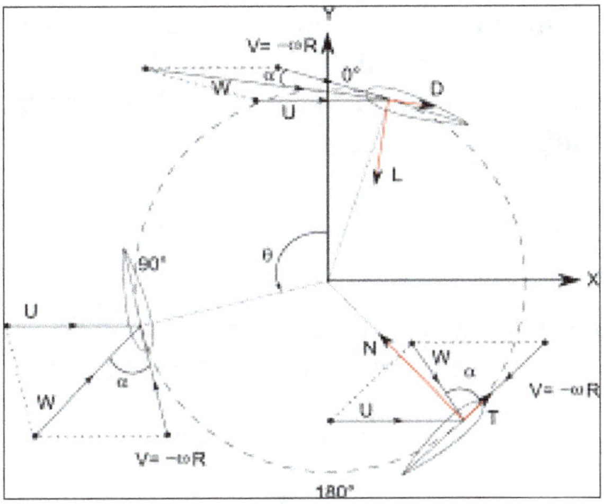


Forces and velocities acting in a Darrieus turbine for various azimuthal positions.

Solving for the relative velocity as the resultant of the tangential and normal components yields:

$$W = \sqrt{V_t^2 + V_n^2}$$

Thus, combining the above with the definitions for the tip speed ratio λ = (ωR)/U yields the following expression for the resultant velocity:

$$W = U\sqrt{1 + 2\lambda\cos\theta + \lambda^2}$$

Angle of attack is solved as:

$$\alpha = \tan^{-1}\left(\frac{V_n}{V_t}\right)$$

Which when substituting the above yields:

$$\alpha = \tan^{-1}\left(\frac{\sin\theta}{\cos\theta + \lambda}\right)$$

The resultant aerodynamic force is resolved either into lift (L) - drag (D) components or normal (N) - tangential (T) components. The forces are considered acting at the quarter-chord point, and the pitching moment is determined to resolve the aerodynamic forces. The aeronautical terms "lift" and "drag" refer to the forces across (lift) and along (drag) the approaching net relative airflow. The tangential force acts along the blade's velocity, pulling the blade around, and the normal force acts radially, pushing against the shaft bearings. The lift and the drag force are useful when dealing with the aerodynamic forces around the blade such as dynamic stall, boundary layer etc.; while when dealing with global performance, fatigue loads, etc., it is more convenient to have a normal-tangential frame. The lift and the drag coefficients are usually normalised by the dynamic pressure of the relative airflow, while the normal and tangential coefficients are usually normalised by the dynamic pressure of undisturbed upstream fluid velocity.

$$C_L = \frac{F_L}{1/2\,\rho A W^2}\ ; \quad C_D = \frac{D}{1/2\,\rho A W^2}\ ; \quad C_T = \frac{T}{1/2\,\rho A U^2 R}\ ; \quad C_N = \frac{N}{1/2\,\rho A U^2}$$

A = Blade Area (not to be confused with the Swept Area, which is equal to the height of the blade/rotor times the rotor diameter), R = Radius of turbine

The amount of power, P, that can be absorbed by a wind turbine:

$$P = \frac{1}{2} C_p \rho A v^3$$

Where $C_p$ is the power coefficient, $\rho$ is air density, $A$ is the swept area of the turbine, and $v$ is the wind speed.

## Advantages

VAWTs offer a number of advantages over traditional Horizontal axis wind turbines (HAWTs):

- Being omni-directional, some forms do not need to track the wind. This means they don't require a complex mechanism and motors to yaw the rotor and pitch the blades.
- VAWTs generally function better than HAWTs in turbulent and gusty winds. HAWTs cannot efficiently harvest such winds, which also cause accelerated fatigue.
- The gearbox of a VAWT takes much less fatigue than that of a HAWT.
- In VAWTs, gearbox replacement and maintenance are simpler and more efficient, as the gearbox is accessible at ground level, so that that no cranes or other large equipment are needed on-site. This reduces costs and impact on the environment. Motor and gearbox failures generally are significant considerations in the operation and maintenance of HAWTs both on and offshore.
- Some designs of VAWTs in suitable situations can use screw pile foundations, which hugely reduces the road transport of concrete and the carbon cost of installation. Screw piles can be fully recycled at the end of their life.
- Wings of the Darrieus type have a constant chord and so are easier to manufacture than the blades of a HAWT, which have a much more complex shape and structure.
- VAWTs can be grouped more closely in wind farms, increasing the generated power per unit of land area.

- VAWTs can be installed on HAWT wind farm below the existing HAWTs; this can supplement the power output of the existing farm.
- Research at Caltech has also shown that a carefully designed wind farm using VAWTs can have an output power ten times that of a HAWT wind farm of the same size.

## Disadvantages

One of the major outstanding challenges facing vertical axis wind turbine technology is dynamic stall of the blades as the angle of attack varies rapidly.

The blades of a VAWT are fatigue-prone due to the wide variation in applied forces during each rotation. This can be overcome by the use of modern composite materials and improvements in design - including the use of aerodynamic wing tips that cause the spreader wing connections to have a static load. The vertically oriented blades can twist and bend during each turn, causing them to break apart.

VAWTs have proven less reliable than HAWTs, although modern designs of VAWTs have overcome many of the issues associated with early designs.

## Applications

The Windspire, a small VAWT intended for individual (home or office) use was developed in the early 2000s by US company Mariah Power. The company reported that several units had been installed across the US by June 2008.

Arborwind, an Ann-Arbor (Michigan, US) based company, produces a patented small VAWT which has been installed at several US locations as of 2013.

In 2011, Sandia National Laboratories wind-energy researchers began a five-year study of applying VAWT design technology to offshore wind farms. The researchers stated: "The economics of offshore windpower are different from land-based turbines, due to installation and operational challenges. VAWTs offer three big advantages that could reduce the cost of wind energy: a lower turbine center of gravity; reduced machine complexity; and better scalability to very large sizes. A lower center of gravity means improved stability afloat and lower gravitational fatigue loads. Additionally, the drivetrain on a VAWT is at or near the surface, potentially making maintenance easier and less time-consuming. Fewer parts, lower fatigue loads and simpler maintenance all lead to reduced maintenance costs."

A 24-unit VAWT demonstration plot was installed in southern California in the early 2010s by Caltech aeronautical professor John Dabiri. His design was incorporated in a 10-unit generating farm installed in 2013 in the Alaskan village of Igiugig.

Dulas, Anglesey received permission in March 2014 to install a prototype VAWT on the breakwater at Port Talbot waterside. The turbine is a new design, supplied by

Wales-based C-FEC (Swansea), and will be operated for a two-year trial. This VAWT incorporates a wind shield which blocks the wind from the advancing blades, and thus requires a wind-direction sensor and a positioning mechanism, as opposed to the "egg-beater" types of VAWTs.

4 Navitas (Blackpool) have been operating two prototype VAWTs since June 2013, powered by a Siemens Power Train, they are due to enter the market in January 2015, with a free technology share to interested parties. Navitas are now in the process of scaling their prototype to 1 MW, (working with PERA Technology) and then floating the turbine on an offshore pontoon. This will reduce the cost of offshore wind energy.

The Dynasphere, is Michael Reynolds' (known for his Earthship house designs) 4th generation vertical axis windmill. These windmills have two 1.5 KW generators and can produce electricity at very low speeds.

## Savonius Wind Turbine

Savonius wind turbine.

Savonius wind turbines are a type of vertical-axis wind turbine (VAWT), used for converting the force of the wind into torque on a rotating shaft. The turbine consists of a number of aerofoils, usually—but not always—vertically mounted on a rotating shaft or framework, either ground stationed or tethered in airborne systems.

The Savonius wind turbine was invented by the Finnish engineer Sigurd Johannes Savonius in 1922. However, Europeans had been experimenting with curved blades on vertical wind turbines for many decades before this. The earliest mention is by the Italian Bishop of Czanad, Fausto Veranzio, who was also an engineer. He wrote in his 1616 book Machinae novae about several vertical axis wind turbines with curved or V-shaped blades. None of his or any other earlier examples reached the state of development made by Savonius. In his Finnish biography there is mention of his intention to develop a turbine-type similar to the Flettner-type, but autorotationary. He experimented with his rotor on small rowing vessels on lakes in his country. No results of his particular investigations are known, but the Magnus effect is confirmed by König. The two Savonius patents: US1697574, were filed in 1925 by Sigurd Johannes Savonius, and US1766765, in 1928.

## Operation.

The Savonius turbine is one of the simplest turbines. Aerodynamically, it is a drag-type device, consisting of two or three scoops. Looking down on the rotor from above, a two-scoop machine would look like an "S" shape in cross section. Because of the curvature, the scoops experience less drag when moving against the wind than when moving with the wind. The differential drag causes the Savonius turbine to spin. Because they are drag-type devices, Savonius turbines extract much less of the wind's power than other similarly-sized lift-type turbines. Much of the swept area of a Savonius rotor may be near the ground, if it has a small mount without an extended post, making the overall energy extraction less effective due to the lower wind speeds found at lower heights.

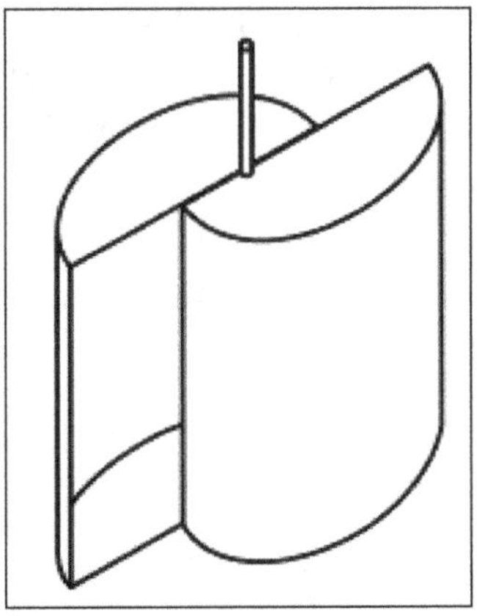

Schematic drawing of a two-scoop Savonius turbine

## Power and Rotational Speed

According to Betz's law, the maximum power that is possible to extract from a rotor is:

$$P_{max} = \frac{16}{27} \rho \cdot r \cdot h \cdot v^3$$

where $\rho$ is the density of air, $h$ and $r$ are the height and radius of the rotor and $v$ is the wind speed. However, in practice the extractable power is about half that (one can argue that only one half of the rotor — the scoop co-moving with the wind — works at each instant of time). Thus, one gets $P_{max} \approx 0.36 \text{kgm}^{-3} \cdot h \cdot r \cdot v^3$.

The angular frequency of a rotor is given by:

$$\omega = \frac{\lambda \cdot v}{r}$$

where $\lambda$ is a dimensionless factor called the tip-speed ratio. $\lambda$ is a characteristic of each specific windmill, and for a Savonius rotor $\lambda$ is typically around unity.

For example, an oil-barrel sized Savonius rotor with $h$ = 1 m and $r$ =0.5 m under a wind of $v$ = 10 m/s, will generate a maximum power of 180 W and an angular speed of 20 rad/s (190 revolutions per minute).

## Use

Savonius turbines are used whenever cost or reliability is much more important than efficiency.

Most anemometers are Savonius turbines for this reason, as efficiency is irrelevant to the application of measuring wind speed. Much larger Savonius turbines have been used to generate electric power on deep-water buoys, which need small amounts of power and get very little maintenance. Design is simplified because, unlike with horizontal axis wind turbines (HAWTs), no pointing mechanism is required to allow for shifting wind direction and the turbine is self-starting. Savonius and other vertical-axis machines are good at pumping water and other high torque, low rpm applications and are not usually connected to electric power grids. In the early 1980s, Risto Joutsiniemi developed a helical rotor version that does not require end plates, has a smoother torque profile and is self-starting in the same way a crossed pair of straight rotors is.

Combined Darrieus–Savonius generator.

The most ubiquitous application of the Savonius wind turbine is the Flettner rotor, which is commonly seen on the roofs of vans and buses and is used as a cooling device. The ventilator was developed by the German aircraft engineer Anton Flettner in the 1920s. It uses the Savonius wind turbine to drive an extractor fan. The vents are still manufactured in the UK by Flettner Ventilator Limited.

Small Savonius wind turbines are sometimes seen used as advertising signs where the rotation helps to draw attention to the item advertised. They sometimes feature a simple two-frame animation.

### Tethered Airborne Savonius Turbines

- Airborne wind turbines.
- Kite types.
- When the Savonius rotor axis is set horizontally and tethered, then kiting results. There are scores of patents and products that use the net lift Magnus effect that occurs in the autorotation of the Savonius rotor. The spin may be mined for some of its energy for making noise, heat, or electricity.

## Small Wind Turbine

A small-scale wind tower.

A group of small wind turbines in a community.

A small wind turbine is a wind turbine used for microgeneration, as opposed to large commercial wind turbines, such as those found in wind farms, with greater individual power output. The Canadian Wind Energy Association (CanWEA) defines "small wind" as ranging from less than 1000 Watt (1 kW) turbines up to 300 kW turbines. The smaller turbines may be as small as a 50 Watt auxiliary power generator for a boat, caravan, or miniature refrigeration unit. The IEC-61400-2:2006 Standard defines small wind turbines as wind turbines with a rotor swept area smaller than 200 m2, generating at a voltage below 1000 Va.c. or 1500 Vd.c.

## Design

Evance R9000 5kW small domestic wind turbine with 5.5m rotor diameter.

Smaller scale turbines for residential scale use are available. Their blades are usually 1.5 to 3.5 metres (4 ft 11 in–11 ft 6 in) in diameter and produce 1-10 kW of electricity at their optimal wind speed. Some units have been designed to be very lightweight in their construction, e.g. 16 kilograms (35 lb), allowing sensitivity to minor wind movements and a rapid response to wind gusts typically found in urban settings and easy mounting much like a television antenna. It is claimed, and a few are certified, as being inaudible even a few feet (about a metre) under the turbine.

The majority of small wind turbines are traditional horizontal axis wind turbines, but vertical axis wind turbines are a growing type of wind turbine in the small-wind market. Makers of vertical axis wind turbines such as WePower, Urban Green Energy, Helix Wind, and Windspire Energy, have reported increasing sales over the previous years.

The generators for small wind turbines usually are three-phase alternating current generators and the trend is to use the induction type. They are options for direct current output for battery charging and power inverters to convert the power back to AC but at constant frequency for grid connectivity. Some models utilize single-phase generators.

Some small wind turbines can be designed to work at low wind speeds, but in general small wind turbines require a minimum wind speed of 4 metres per second (13 ft/s).

Dynamic braking regulates the speed by dumping excess energy, so that the turbine continues to produce electricity even in high winds. The dynamic braking resistor may be installed inside the building to provide heat (during high winds when more heat is lost by the building, while more heat is also produced by the braking resistor). The location makes low voltage (around 12 volt) distribution practical.

Small units often have direct drive generators, direct current output, lifetime bearings and use a vane to point into the wind. Larger, more costly turbines generally have geared power trains, alternating current output and are actively pointed into the wind. Direct drive generators are also used on some large wind turbines.

## Materials

While natural fibers face quality variations, high moisture uptake and low thermal stability that make them undesirable for larger blades experiencing high amounts of stress, the lower stress small turbines used in rural electrification and small-scale renewable systems can still take advantage of them. Hemp, flax, wood and bamboo are all candidate blade materials for small turbines. Nepal has used small blade turbines made of coated timber and, of the available wood materials including Sal, Saur, Sisau, Uttish, Tuni and Okhar, pine and lakuri wood were identified as performing the best based on their ready availability, cost and growth time average density, high stiffness, and breaking strain. Coatings are also generally used to reduce moisture and white enamel with primer has been found to be particularly effective. Sitka spruce, (used in propellers) and Douglas Fir have also been used in turbine blades.

Beyond wood, bamboo-based composites may also be used in both large and small wind turbines due to their low density and carbon sequestration ability—which makes bamboo materials environmentally friendly. Furthermore, relative to wood, bamboo has higher facture toughness, higher strength, lower processing costs and fast growth rate. Ongoing materials developments include bamboo laminates using resins and hybrid bamboo carbon-fiber materials.

A range of synthetic materials including carbon fiber reinforced polymers, nanocomposites, and E-glass-polyester have also been used.

## Installation

Turbines are often mounted on a tower to raise them above any nearby obstacles. One rule of thumb is that turbines should be at least 9 m (30 ft) higher than anything within 150 m (490 ft). Better locations for wind turbines are far away from large upwind obstacles. Measurements made in a boundary layer wind tunnel have indicated that significant detrimental effects associated with nearby obstacles can extend up to 80 times the obstacle's height downwind. However, this is an extreme case. Another approach to siting a small turbine is to use a shelter model to predict how nearby obstacles will affect local wind conditions. Models of this type are general and can be applied to any site. They are often developed based on actual wind measurements, and can estimate flow properties such as mean wind speed and turbulence levels at a potential turbine location, taking into account the size, shape, and distance to any nearby obstacles.

A small wind turbine can be installed on a roof. Installation issues then include the strength of the roof, vibration, and the turbulence caused by the roof ledge. Small-scale rooftop turbines suffer from turbulence and rarely generate significant amounts of power, especially in towns and cities.

Twisted savonius wind turbine.

## Unconventional Wind Turbines

Unconventional wind turbines are those that differ significantly from the most common types in use.

As of 2012, the most common type of wind turbine is the three-bladed upwind Horizontal axis wind turbine (HAWT), where the turbine rotor is at the front of the nacelle and facing the wind upstream of its supporting turbine tower. A second major unit type is the vertical-axis wind turbine (VAWT), with blades extending upwards, supported by a rotating framework.

Due to the large growth of the wind power industry, many wind turbine designs exist, are in development, or have been proposed. The variety of designs reflects ongoing commercial, technological, and inventive interests in harvesting wind resources more efficiently and in greater volume.

Some unconventional designs have entered commercial use, while others have only been demonstrated or are only theoretical concepts. Unconventional designs cover a wide gamut of innovations, including different rotor types, basic functionalities, supporting structures and form-factors.

Crosswind kite generator with fast motion transfer.

## Horizontal Axis

### Twin-bladed Rotor

Nearly all modern wind turbines use rotors with three blades, but some use only two blades. This was the type used at Kaiser-Wilhelm-Koog, Germany, where a large experimental two-bladed unit—the GROWIAN, or Große Windkraftanlage (big wind turbine)—operated from 1983 to 1987. Other prototypes and wind turbine types were manufactured by NedWind. The Eemmeerdijk Wind Park in Zeewolde, Netherlands uses only two-bladed turbines. Wind turbines with two blades are manufactured by Windflow Technology, Mingyang Wind Power, GC China Turbine Corp and Nordic Windpower,. The NASA wind turbines each had 2-blade rotors, producing the same energy at lower cost than three-blade rotor designs.

### Downwind Rotor

Nearly all wind turbines place the rotor in front of the nacelle when the wind is blowing (upwind design). Some turbines place the rotor behind the nacelle (downwind design). This design has the advantage that the turbine can be made to passively align itself with the wind, reducing cost. The main drawback is that the load on the blades changes as they pass behind the tower, increasing fatigue loading, and potentially exciting resonances in other turbine structures.

### Ducted Rotor

A research project, the ducted rotor consists of a turbine inside a duct that flares at the back. They are also referred as Diffuser-Augmented Wind Turbines (i.e. DAWT). Its main advantage is that it can operate in a wide range of winds and generate a higher power per unit of rotor area. Another advantage is that the generator operates at a high rotation rate, so it doesn't require a bulky gearbox, allowing the mechanical portion to be smaller and lighter. A disadvantage is that (apart from the gearbox) it is more complicated than the unducted rotor and the duct's weight increases tower weight. The Éolienne Bollée is an example of a DAWT.

### Co-axial and Multi-rotor

Two or more rotors may be mounted to a single driveshaft, with their combined co-rotation together turning the same generator: fresh wind is brought to each rotor by sufficient spacing between rotors combined with an offset angle (alpha) from the wind direction. Wake vorticity is recovered as the top of a wake hits the bottom of the next rotor. Power was multiplied several times using co-axial, multiple rotors in testing conducted by inventor and researcher Douglas Selsam in 2004. The first commercially available co-axial multi-rotor turbine is the patented dual-rotor American Twin Superturbine from Selsam Innovations in California, with 2 propellers separated by 12 feet. It is the most powerful 7-foot-diameter

(2.1 m) turbine available, due to this extra rotor. In 2015, Iowa State University aerospace engineers Hui Hu and Anupam Sharma were optimizing designs of multi-rotor systems, including a Horizontal axis co-axial dual-rotor model. In addition to a conventional three-blade rotor, it has a smaller secondary three-blade rotor, covering the near-axis region usually inefficiently harvested. Preliminary results indicated 10-20% gains, less efficient than is claimed by existing counter-rotating designs.

Counter-rotating wind turbine (dual rotor).

## Counter-rotating Horizontal Axis

When a system expels or accelerates mass in one direction, the accelerated mass causes a proportional but opposite force on that system. The spinning blade of a single rotor wind turbine causes a significant amount of tangential or rotational air flow. The energy of this tangential air flow is wasted in a single-rotor propeller design. To use this wasted effort, the placement of a second rotor behind the first takes advantage of the disturbed airflow, and can gain up to 40% more energy from a given swept area as compared with a single rotor. Other advantages of contra-rotation include no gear boxes and auto-centering on the wind (no yaw motors/mechanism required). A patent application dated 1992 exists based on work done with the Trimblemill.

When the counter-rotating turbines are on the same side of the tower, the blades in front are angled forwards slightly so as to avoid hitting the rear ones. If the turbine blades are on opposite sides of the tower, it is best that the blades at the back be smaller than the blades at the front and set to stall at a higher wind speed. This allows the generator to function at a wider wind speed range than a single-turbine generator for a given tower. To reduce sympathetic vibrations, the two turbines should turn at speeds with few common multiples, for example 7:3 speed ratio.

When land or sea area for a second wind turbine does not come at a premium the 40% gain with a second rotor has to be compared with a 100% gain via the expense of a separate foundation and tower with cabling for the second turbine. As of 2005, no large, counter-rotating HAWTs are commercially sold.

## Furling Tail and Twisting Blades

In addition to variable pitch blades, furling tails and twisting blades are other improvements on wind turbines. Similar to the variable pitch blades, they may also greatly increase efficiency and be used in "do-it-yourself" construction.

## Windmill Style

De Nolet is a wind turbine in Rotterdam disguised as a windmill.

## Bladeless

## Boundary Layer

The boundary layer or Tesla turbine uses boundary layers instead of blades.

One modern version is the Fuller turbine. The concept is similar to a stack of disks on a central shaft, separated by a small air gap. The surface tension of air in the small gaps creates friction, rotating the disks around the shaft. Vanes direct the air for improved performance, hence it is not strictly bladeless.

## Vaneless Ion Wind Generator

A vaneless ion wind generator is a theoretical device that produces electrical energy by using the wind to move electric charge from one electrode to another.

## Piezoelectric

Piezoelectric wind turbines work by flexing piezoelectric crystals as they rotate, sufficient to power small electronic devices. They operate with diameters on the scale of 10 centimeters.

## Solar Updraft Tower

Wind turbines may be used in conjunction with a solar collector to extract energy from air heated by the sun and rising through a large vertical updraft tower.

## Vortex

The Vortex Bladeless device maximizes vortex shedding, using the vorticity in wind to flutter a lightweight vertical pole, which delivers that energy to a generator at the bottom of the pole. The design has been criticized for its efficiency of 40%, compared to 70% for conventional designs. However, individual poles can be placed more closely together, offsetting the losses. The design avoids mechanical components, lowering costs. The system also does not threaten bird life and operates silently.

### Saphonian

The Saphonian design uses an oscillating dish to drive a piston, which then connects to a generator.

### Windbeam

The Windbeam generator consists of a beam suspended by springs within an outer frame. The beam oscillates rapidly when exposed to airflow due to multiple fluid flow phenomena. A linear alternator converts the beam motion. The absence of bearings and gears eliminates frictional inefficiencies and noise. The generator can operate in low-light environments unsuitable for solar panels (e.g. HVAC ducts). Costs are low due to low cost components and simple construction.

### Wind Belt

Windbelt is a flexible, tensioned belt that vibrates from the passing flow of air, due to aeroelastic flutter. A magnet, mounted at one end of the belt oscillates in and out of coiled windings, producing electricity. The inventor is Shawn Frayne.

### Aerial

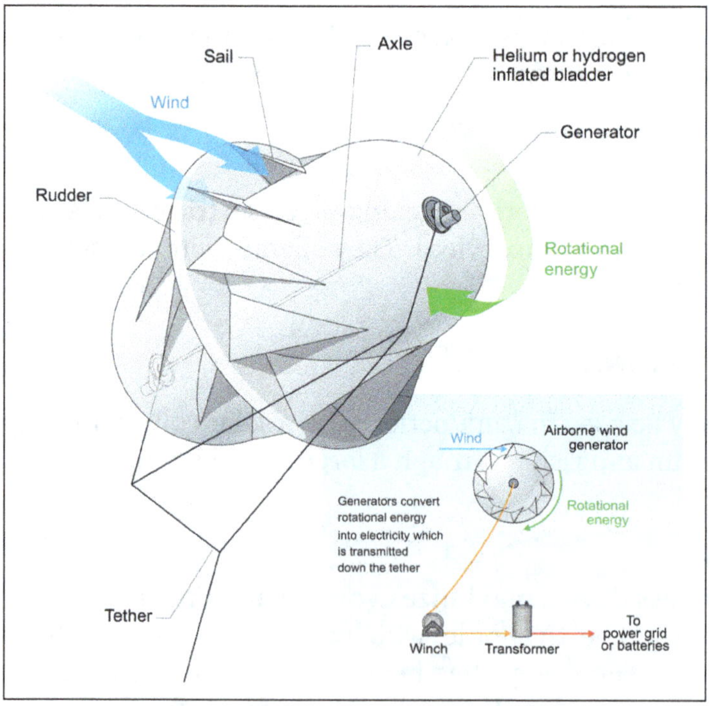


Concept for an airborne wind generator.

Wind turbines could be flown in high-speed winds using high altitude wind power tactics, taking advantage of high altitude winds.

A system of tethered kites could capture energy from high-altitude winds.

## Vertical

### Gorlov

The Gorlov helical turbine (GHT) is a modification of the Darrieus turbine design that uses helical blades/foils.

### Enclosed Blades

One design uses many nylon blades to run a generator. Its permanent magnets are on the tips of the blades, while the stator is a ring outside the blades.

### H-rotor

The giromill is a vertical axis turbine that rotates one blade in one direction while another moves in the opposite direction. Consequently, only one blade is working at a time. Its efficiency is low.

### O-Wind Turbine

A omnidirectional turbine which uses the Bernoulli principle to generate energy using wind from any direction. The design is spherical with a number of ducts across the surface, a pressure difference causes the rotation. The design won the James Dyson Award 2018.

## Components

### INVELOX

SheerWind's INVELOX technology was developed by Dr. Daryoush Allaei. The invention captures and delivers wind to a turbine. In a sense, INVELOX is a wind injection system, much like a fuel injection system for cars. It works by accelerating the wind. A large intake captures wind and funnels it to a concentrator that ends in a Venturi section and finally wind exits from a diffuser. Turbine(s) are placed inside the Venturi section of the INVELOX. Inside the Venturi the dynamic pressure is high while the static pressure is low. The Turbine converts dynamic pressure or kinetic energy to mechanical rotation and thereby to electrical power using a generator. The device has been constructed and tested, but has been criticized for lack of efficiency. As of 2017, prototypes are being installed.

## Applications

### Rooftop

Wind-turbines can be installed on building roofs. Examples include Marthalen Landi-Silo in Switzerland, Council House 2 in Melbourne, Australia. Ridgeblade in the

UK is a vertical wind turbine on its side mounted on the apex of a pitched roof. Another example installed in France is the Aeolta AeroCube. Discovery Tower is an office building in Houston, Texas, that incorporates 10 wind turbines.

The Museum of Science in Boston, Massachusetts began constructing a rooftop Wind Turbine Lab in 2009. The lab is testing nine wind turbines from five different manufacturers. Rooftop wind turbines may suffer from turbulence, especially in cities, which reduces power output and accelerates turbine wear. The lab seeks to address the general lack of performance data for urban wind turbines.

Due to structural limitations of buildings, limited space in urban areas, and safety considerations, building turbines are usually small (with capacities in the low kilowatts). An exception is the Bahrain World Trade Centre with three 225 kW wind turbines mounted between twin skyscrapers.

### Traffic-driven

Proposals call for generating power from the energy in the draft created by traffic.

## Darrieus Wind Turbine

A Darrieus is a type of vertical axis wind turbine (VAWT) generator. Unlike the Savonius wind turbine, the Darrieus is a lift-type VAWT. Rather than collecting the wind in cups dragging the turbine around, a Darrieus uses lift forces generated by the wind hitting aerofoils to create rotation.

### Benefits of the Darrieus Wind Turbine

A Darrieus wind turbine can spin at many times the speed of the wind hitting it (i.e. the tip speed ratio(TSR) is greater than 1). Hence a Darrieus wind turbine generates less torque than a Savonius but it rotates much faster. This makes Darrieus wind turbines much better suited to electricity generation rather than water pumping and similar activities. The centrifugal forces generated by a Darrieus turbine are very large and act on the turbine blades which therefore have to be very strong – however the forces on the bearings and generator are usually lower than are the case with a Savonius.

## Self Starting Darrieus Wind Turbines

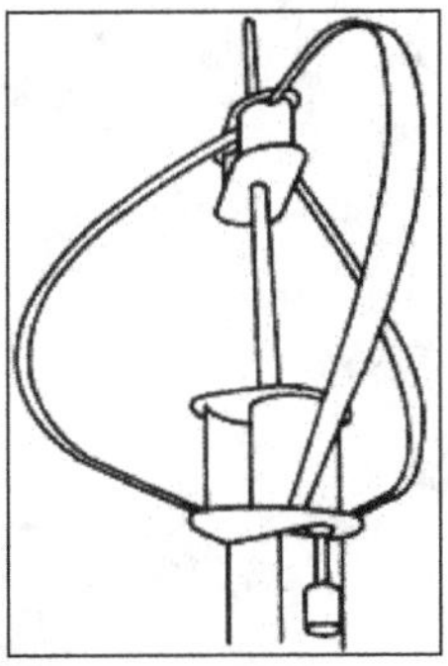

Darrieus wind turbines are not self-starting. Therefore a small powered motor is required to start off the rotation, and then when it has enough speed the wind passing across the aerofoils starts to generate torque and the rotor is driven around by the wind. An alternative is shown in the illustration above. Two small Savonius rotors are mounted on the shaft of the Darrieus turbine to start rotation. These slow down the Darrieus turbine when it gets going however they make the whole device a lot simpler and easier to maintain.

Variants of the Darrieus wind turbine are the giromill and cycloturbine.

## References


- Wind-energy: openei.org, Retrieved 14 June, 2020
- M. Lange and U. Focken. Physical approach to short-term wind power forecast, Springer, ISBN 3-540-25662-8, 2005
- Navid Goudarzi (June 2013). "A Review on the Development of the Wind Turbine Generators across the World". International Journal of Dynamics and Control. 1 (2): 192–202. doi:10.1007/s40435-013-0016-y
- Darrieus-wind-turbines, wind: reuk.co.uk, Retrieved 23 March, 2020
- Goldstein, Leo (2013). "Theoretical analysis of an airborne wind energy conversion system with a ground generator and fast motion transfer". Energy. 55: 987–995. doi:10.1016/j.energy.2013.03.087
- Gipe, Paul. Wind energy basics: a guide to home- and community-scale wind energy systems. Chelsea Green Publishing, 2009. Accessed: 18 December 2010. ISBN 1-60358-030-1 ISBN 978-1-60358-030-4

# Solar Energy 3

- **Solar Cell**
- **Organic Solar Cell**
- **Solar Panel**
- **Floating Solar**
- **Photovoltaics**

Solar energy is the radiant light and heat from the sun that is utilized for generating electricity or causing chemical reactions. It use a range of technologies such as solar heating, solar thermal energy, photovoltaics, solar architecture, etc. This chapter closely examines solar energy to provide an extensive understanding of the subject.

Solar energy is radiation from the Sun capable of producing heat, causing chemical reactions, or generating electricity. The total amount of solar energy incident on Earth is vastly in excess of the world's current and anticipated energy requirements. If suitably harnessed, this highly diffused source has the potential to satisfy all future energy needs. In the 21st century solar energy is expected to become increasingly attractive as a renewable energy source because of its inexhaustible supply and its nonpolluting character, in stark contrast to the finite fossil fuels coal, petroleum, and natural gas.

Solar panel.

The Sun is an extremely powerful energy source, and sunlight is by far the largest source of energy received by Earth, but its intensity at Earth's surface is actually quite low. This is essentially because of the enormous radial spreading of radiation from the distant Sun. A relatively minor additional loss is due to Earth's atmosphere and clouds, which absorb or scatter as much as 54 percent of the incoming sunlight. The sunlight that reaches the ground consists of nearly 50 percent visible light, 45 percent infrared radiation, and smaller amounts of ultraviolet and other forms of electromagnetic radiation.

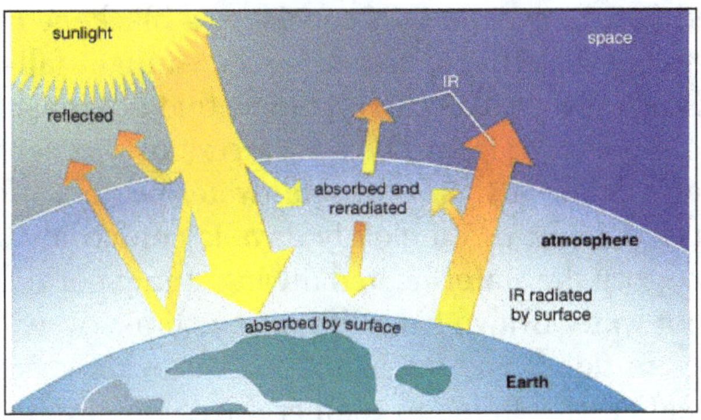


Solar energy.

Reflection and absorption of solar energy. Although some incoming sunlight is reflected by Earth's atmosphere and surface, most is absorbed by the surface, which is warmed.

## Uses of Solar Energy

The potential for solar energy is enormous, since about 200,000 times the world's total daily electric-generating capacity is received by Earth every day in the form of solar energy. Unfortunately, though solar energy itself is free, the high cost of its collection, conversion, and storage still limits its exploitation in many places. Solar radiation can be converted either into thermal energy (heat) or into electrical energy, though the former is easier to accomplish.

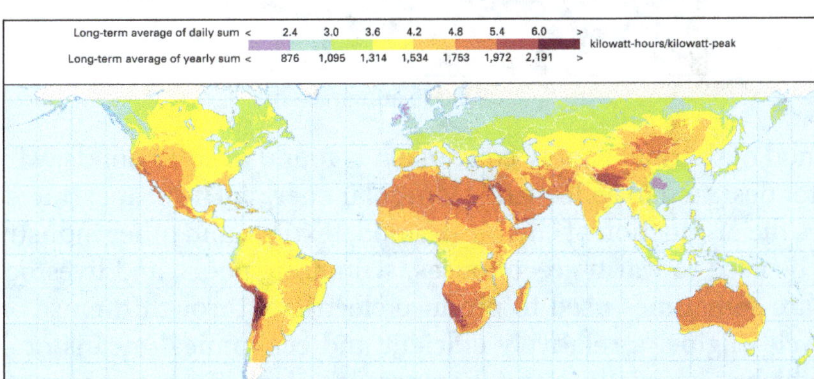


Earth's photovoltaic power potential.

## Thermal Energy

Among the most common devices used to capture solar energy and convert it to thermal energy are flat-plate collectors, which are used for solar heating applications. Because the intensity of solar radiation at Earth's surface is so low, these collectors must be large in area. Even in sunny parts of the world's temperate regions, for instance, a collector must have a surface area of about 40 square metres (430 square feet) to gather enough energy to serve the energy needs of one person.

The most widely used flat-plate collectors consist of a blackened metal plate, covered with one or two sheets of glass, that is heated by the sunlight falling on it. This heat is then transferred to air or water, called carrier fluids, that flow past the back of the plate. The heat may be used directly, or it may be transferred to another medium for storage. Flat-plate collectors are commonly used for solar water heaters and house heating. The storage of heat for use at night or on cloudy days is commonly accomplished by using insulated tanks to store the water heated during sunny periods. Such a system can supply a home with hot water drawn from the storage tank, or, with the warmed water flowing through tubes in floors and ceilings, it can provide space heating. Flat-plate collectors typically heat carrier fluids to temperatures ranging from 66 to 93 °C (150 to 200 °F). The efficiency of such collectors (i.e., the proportion of the energy received that they convert into usable energy) ranges from 20 to 80 percent, depending on the design of the collector.

Solar heating.

Another method of thermal energy conversion is found in solar ponds, which are bodies of salt water designed to collect and store solar energy. The heat extracted from such ponds enables the production of chemicals, food, textiles, and other industrial products and can also be used to warm greenhouses, swimming pools, and livestock buildings. Solar ponds are sometimes used to produce electricity through the use of the organic Rankine cycle engine, a relatively efficient and economical means of solar energy conversion, which is especially useful in remote locations. Solar ponds are fairly expensive to install and maintain and are generally limited to warm rural areas.

On a smaller scale, the Sun's energy can also be harnessed to cook food in specially designed solar ovens. Solar ovens typically concentrate sunlight from over a wide area to a central point, where a black-surfaced vessel converts the sunlight into heat. The ovens are typically portable and require no other fuel inputs.

A monk using a solar-powered cookstove.

## Electricity Generation

Solar radiation may be converted directly into electricity by solar cells (photovoltaic cells). In such cells, a small electric voltage is generated when light strikes the junction between a metal and a semiconductor (such as silicon) or the junction between two different semiconductors. The power generated by a single photovoltaic cell is typically only about two watts. By connecting large numbers of individual cells together, however, as in solar-panel arrays, hundreds or even thousands of kilowatts of electric power can be generated in a solar electric plant or in a large household array. The energy efficiency of most present-day photovoltaic cells is only about 15 to 20 percent, and, since the intensity of solar radiation is low to begin with, large and costly assemblies of such cells are required to produce even moderate amounts of power.

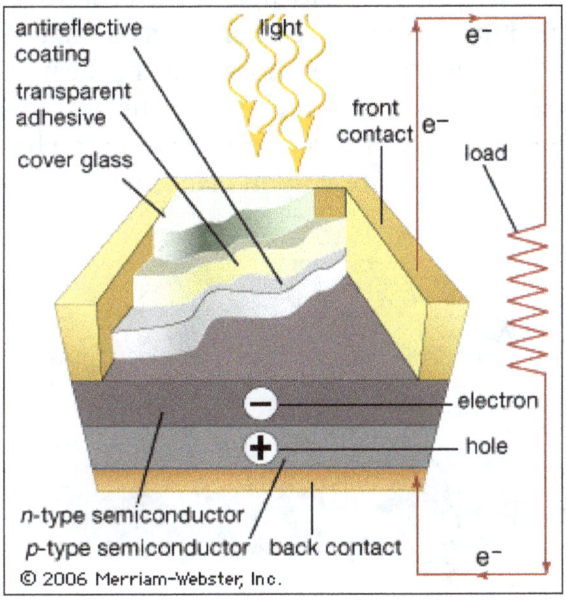

When sunlight strikes a solar cell, an electron is freed by the photoelectric effect. The two dissimilar semiconductors possess a natural difference in electric potential (voltage), which causes the electrons to flow through the external circuit, supplying power to the load. The flow of electricity results from the characteristics of the semiconductors and is powered entirely by light striking the cell.

Small photovoltaic cells that operate on sunlight or artificial light have found major use in low-power applications—as power sources for calculators and watches, for example. Larger units have been used to provide power for water pumps and communications systems in remote areas and for weather and communications satellites. Classic crystalline silicon panels and emerging technologies using Thin film solar cells, including building-integrated photovoltaics, can be installed by homeowners and businesses on their rooftops to replace or augment the conventional electric supply.

Solar power.

Concentrated solar-power plant.

Concentrated solar power plants employ concentrating, or focusing, collectors to concentrate sunlight received from a wide area onto a small blackened receiver, thereby considerably increasing the light's intensity in order to produce high temperatures. The arrays of carefully aligned mirrors or lenses can focus enough sunlight to heat a target to temperatures of 2,000 °C (3,600 °F) or more. This heat can then be used to operate a boiler, which in turn generates steam for a steam turbine electric generator power plant. For producing steam directly, the movable mirrors can be arranged so as to concentrate large amounts of solar radiation upon blackened pipes through which water is circulated and thereby heated.

Solar reflector.

## Other Applications

Solar energy is also used on a small scale for purposes other than those described above. In some countries, for instance, solar energy is used to produce salt from seawater by evaporation. Similarly, solar-powered desalination units transform salt water into drinking water by converting the Sun's energy to heat, directly or indirectly, to drive the desalination process.

Solar technology has also emerged for the clean and renewable production of hydrogen as an alternative energy source. Mimicking the process of photosynthesis, artificial leaves are silicon-based devices that use solar energy to split water into hydrogen and oxygen, leaving virtually no pollutants. Further work is needed to improve the efficiency and cost-effectiveness of these devices for industrial use.

## Benefits and Drawbacks

### Sustainable

The advantage of solar energy is that it is a sustainable alternative to fossil fuels. While fossil fuels have an expiration date that may be fast approaching, the sun is likely to be around for at least a few billion years.

### Low Environmental Impact

Solar energy has a substantially reduced impact on the environment compared to fossil fuels. Its greenhouse gas emissions are inconsequential as the technology does not require any fuel combustion. Also, although concentrating solar thermal plants (CSP) are comparatively inefficient in their water usage depending on the type of technology being used, the right technology significantly increases efficiency while photovoltaic (PV) solar cells do not require any water when generating electricity.

### Energy Independence

Since the sun shines across the globe, it makes every country a potential energy producer, thus allowing for greater energy independence and security. Solar energy doesn't only promise to bring security and independence at the national level; solar panels can be installed on individual homes, providing power that does not depend on being connected to a larger electrical grid.

## The Drawbacks of Solar Energy

### Intermittency

One of the biggest problems that solar energy technology poses is that energy is only generated while the sun is shining. That means nighttime and overcast days can interrupt the supply. The shortage created by this interruption would not be a problem if there were low-cost ways of storing energy as extremely sunny periods can actually generate excess capacity. In fact, Germany—a leader in solar energy technology—is now focusing on developing adequate energy storage to deal with this issue.

### Land Use

Another concern is that solar energy may take up a significant amount of land and cause land degradation or habitat loss for wildlife. While solar PV systems can be fixed to already existing structures, larger utility-scale PV systems may require up to 3.5 to 10 acres per megawatt and CSP facilities require anywhere from 4 to 16.5 acres per megawatt. However, the impact can be reduced by placing facilities in low-quality areas or along existing transportation and transmission corridors.

### Scarcity of Materials

Certain solar technologies require rare materials in their production. This, however, is primarily a problem for PV technology rather than CSP technology. Also, it is not so much a lack of known reserves as much as it is the inability of current production to meet future demand: Many of the rare materials are byproducts of other processes rather than the focus of targeted mining efforts. Recycling PV material and advances in nanotechnology that increase solar-cell efficiency could both help boost supply, but perhaps finding material substitutes that exist in greater abundance could play a role.

### Environmental Downside

The one environmental downside to solar technology is that it contains many of the same hazardous materials as electronics. As solar becomes a more popular energy, the problem of disposing the hazardous waste becomes an additional challenge. However, assuming the challenge of proper disposal is met, the reduced greenhouse gas emissions that solar energy offers makes it an attractive alternative to fossil fuels.

## Solar Cell

Solar cell, also called photovoltaic cell, is any device that directly converts the energy of light into electrical energy through the photovoltaic effect. The overwhelming majority of solar cells are fabricated from silicon—with increasing efficiency and lowering cost as the materials range from amorphous (noncrystalline) to polycrystalline to crystalline (single crystal) silicon forms. Unlike batteries or fuel cells, solar cells do not utilize chemical reactions or require fuel to produce electric power, and, unlike electric generators, they do not have any moving parts.

Solar cells can be arranged into large groupings called arrays. These arrays, composed of many thousands of individual cells, can function as central electric power stations, converting sunlight into electrical energy for distribution to industrial, commercial, and residential users. Solar cells in much smaller configurations, commonly referred to as solar cell panels or simply solar panels, have been installed by homeowners on their rooftops to replace or augment their conventional electric supply. Solar cell panels also are used to provide electric power in many remote terrestrial locations where conventional electric power sources are either unavailable or prohibitively expensive to install. Because they have no moving parts that could need maintenance or fuels that would require replenishment, solar cells provide power for most space installations, from communications and weather satellites to space stations. (Solar power is insufficient for space probes sent to the outer planets of the solar system or into interstellar space, however, because of the diffusion of radiant energy with distance from the Sun.) Solar cells have also been used in consumer products, such as electronic toys, handheld calculators, and portable radios. Solar cells used in devices of this kind may utilize artificial light (e.g., from incandescent and fluorescent lamps) as well as sunlight.

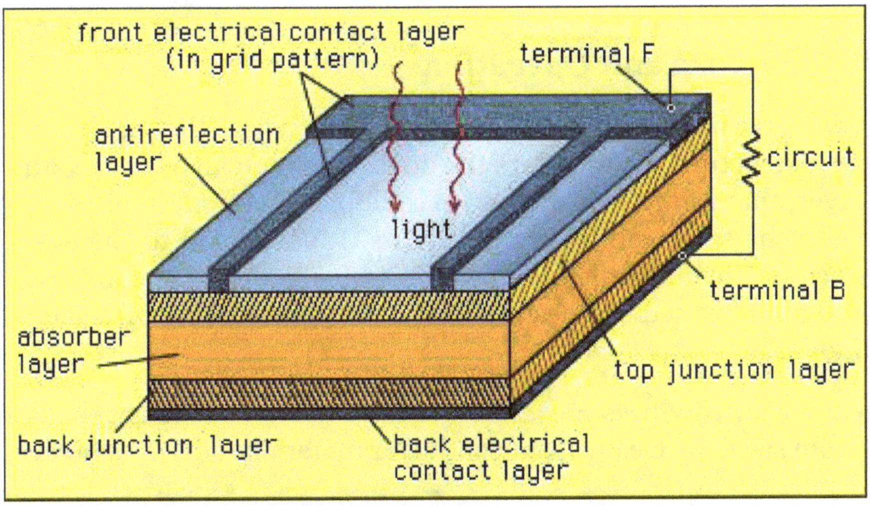


A commonly used solar cell structure. In many such cells, the absorber layer and the back junction layer are both made of the same material.

The International Space Station (ISS) was built in sections beginning in 1998. By December 2000 the major elements of the partially completed station included the American-built connecting node Unity and two Russian-built units—Zarya, a power module, and Zvezda, the initial living quarters. A Russian spacecraft, which carried up the station's first three-person crew, is docked at the end of Zvezda. The photograph was taken from the space shuttle Endeavour.

While total photovoltaic energy production is minuscule, it is likely to increase as fossil fuel resources shrink. In fact, calculations based on the world's projected energy consumption by 2030 suggest that global energy demands would be fulfilled by solar panels operating at 20 percent efficiency and covering only about 496,805 square km (191,817 square miles) of Earth's surface. The material requirements would be enormous but feasible, as silicon is the second most abundant element in Earth's crust. These factors have led solar proponents to envision a future "solar economy" in which practically all of humanity's energy requirements are satisfied by cheap, clean, renew-able sunlight.

International Space Station.

## Solar Cell Structure and Operation

Solar cells, whether used in a central power station, a satellite, or a calculator, have the same basic structure. Light enters the device through an optical coating, or antireflection layer, that minimizes the loss of light by reflection; it effectively traps the light falling on the solar cell by promoting its transmission to the energy-conversion layers below. The antireflection layer is typically an oxide of silicon, tantalum, or titanium that is formed on the cell surface by spin-coating or a vacuum deposition technique.

A solar energy plant produces megawatts of electricity. Voltage is generated by solar cells made from specially treated semiconductor materials, such as silicon.

Solar energy; solar cell.

The three energy-conversion layers below the antireflection layer are the top junction layer, the absorber layer, which constitutes the core of the device, and the back junction layer. Two additional electrical contact layers are needed to carry the electric current out to an external load and back into the cell, thus completing an electric circuit. The electrical contact layer on the face of the cell where light enters is generally present in some grid pattern and is composed of a good conductor such as a metal. Since metal blocks light, the grid lines are as thin and widely spaced as is possible without impairing collection of the current produced by the cell. The back electrical contact layer has no such diametrically opposed restrictions. It need simply function as an electrical contact and thus covers the entire back surface of the cell structure. Because the back layer also must be a very good electrical conductor, it is always made of metal.

Since most of the energy in sunlight and artificial light is in the visible range of electromagnetic radiation, a solar cell absorber should be efficient in absorbing radiation at those wavelengths. Materials that strongly absorb visible radiation belong to a class of substances known as semiconductors. Semiconductors in thicknesses of about one-hundredth of a centimetre or less can absorb all incident visible light; since the junction-forming and contact layers are much thinner, the thickness of a solar cell is essentially that of the absorber. Examples of semiconductor materials employed in solar cells include silicon, gallium arsenide, indium phosphide, and copper indium selenide.

When light falls on a solar cell, electrons in the absorber layer are excited from a lower-energy "ground state," in which they are bound to specific atoms in the solid, to a higher "excited state," in which they can move through the solid. In the absence of the junction-forming layers, these "free" electrons are in random motion, and so there can be no oriented direct current. The addition of junction-forming layers, however, induces a built-in electric field that produces the photovoltaic effect. In effect, the electric field gives a collective motion to the electrons that flow past the electrical contact layers into an external circuit where they can do useful work.

The materials used for the two junction-forming layers must be dissimilar to the absorber in order to produce the built-in electric field and to carry the electric current. Hence, these may be different semiconductors (or the same semiconductor with different types of conduction), or they may be a metal and a semiconductor. The materials used to construct the various layers of solar cells are essentially the same as those used to produce the diodes and transistors of solid-state electronics and microelectronics. Solar cells and microelectronic devices share the same basic technology. In solar cell fabrication, however, one seeks to construct a large-area device because the power produced is proportional to the illuminated area. In microelectronics the goal is, of course, to construct electronic components of ever smaller dimensions in order to increase their density and operating speed within semiconductor chips, or integrated circuits.

The photovoltaic process bears certain similarities to photosynthesis, the process by which the energy in light is converted into chemical energy in plants. Since solar cells obviously cannot produce electric power in the dark, part of the energy they develop under light is stored, in many applications, for use when light is not available. One common means of storing this electrical energy is by charging electrochemical storage batteries. This sequence of converting the energy in light into the energy of excited electrons and then into stored chemical energy is strikingly similar to the process of photosynthesis.

## Development of Solar Cells

The development of solar cell technology stems from the work of French physicist Antoine-César Becquerel in 1839. Becquerel discovered the photovoltaic effect while experimenting with a solid electrode in an electrolyte solution; he observed that voltage developed when light fell upon the electrode. About 50 years later, Charles Fritts constructed the first true solar cells using junctions formed by coating the semiconductor selenium with an ultrathin, nearly transparent layer of gold. Fritts's devices were very inefficient converters of energy; they transformed less than 1 percent of absorbed light energy into electrical energy. Though inefficient by today's standards, these early solar cells fostered among some a vision of abundant, clean power. In 1891 R. Appleyard wrote of the blessed vision of the Sun, no longer pouring his energies unrequited into space, but by means of photo-electric cells, these powers gathered into electrical storehouses to the total extinction of steam engines, and the utter repression of smoke.

By 1927 another metal-semiconductor-junction solar cell, in this case made of copper and the semiconductor copper oxide, had been demonstrated. By the 1930s both the selenium cell and the copper oxide cell were being employed in light-sensitive devices, such as photometers, for use in photography. These early solar cells, however, still had energy-conversion efficiencies of less than 1 percent. This impasse was finally overcome with the development of the silicon solar cell by Russell Ohl in 1941. Thirteen years later, aided by the rapid commercialization of silicon technology needed to fabricate the transistor, three other American researchers—Gerald Pearson, Daryl Chapin, and Calvin Fuller—demonstrated a silicon solar cell capable of a 6 percent energy-conversion efficiency when used in direct sunlight. By the late 1980s silicon cells, as well as cells made of gallium arsenide, with efficiencies of more than 20 percent had been fabricated. In 1989 a concentrator solar cell in which sunlight was concentrated onto the cell surface by means of lenses achieved an efficiency of 37 percent owing to the increased intensity of the collected energy. By connecting cells of different semiconductors optically and electrically in series, even higher efficiencies are possible, but at increased cost and added complexity. In general, solar cells of widely varying efficiencies and cost are now available.

# Organic Solar Cell

An organic solar cell (OSC) or plastic solar cell is a type of photovoltaic that uses organic electronics, a branch of electronics that deals with conductive organic polymers or small organic molecules, for light absorption and charge transport to produce electricity from sunlight by the photovoltaic effect. Most organic photovoltaic cells are polymer solar cells.

Schematic of plastic solar cells.

The molecules used in organic solar cells are solution-processable at high throughput and are cheap, resulting in low production costs to fabricate a large volume. Combined with the flexibility of organic molecules, organic solar cells are potentially cost-effective for photovoltaic applications. Molecular engineering (e.g. changing the length and functional group of polymers) can change the band gap, allowing for electronic tunability. The optical absorption coefficient of organic molecules is high, so a large amount of light can be absorbed with a small amount of materials, usually on the order of hundreds of nanometers. The main disadvantages associated with organic photovoltaic cells are low efficiency, low stability and low strength compared to inorganic photovoltaic cells such as silicon solar cells.

Organic Photovoltaic.

Compared to silicon-based devices, polymer solar cells are lightweight (which is important for small autonomous sensors), potentially disposable and inexpensive to fabricate (sometimes using printed electronics), flexible, customizable on the molecular level and potentially have less adverse environmental impact. Polymer solar cells also have the potential to exhibit transparency, suggesting applications in windows, walls, flexible electronics, etc. An example device is shown in figure. The disadvantages of polymer solar cells are also serious: they offer about 1/3 of the efficiency of hard materials, and experience substantial photochemical degradation.

Polymer solar cells inefficiency and stability problems, combined with their promise of low costs and increased efficiency made them a popular field in solar cell research. As of 2015, polymer solar cells were able to achieve over 10% efficiency via a tandem structure. In 2018, a record breaking efficiency for organic photovoltaics of 17.3% was reached via tandem structure.

## Physics

A photovoltaic cell is a specialized semiconductor diode that converts light into direct current (DC) electricity. Depending on the band gap of the light-absorbing material, photovoltaic cells can also convert low-energy, infrared (IR) or high-energy, ultraviolet (UV) photons into DC electricity. A common characteristic of both the small molecules and polymers used as the light-absorbing material in photovoltaics is that they all have large conjugated systems. A conjugated system is formed where carbon atoms covalently bond with alternating single and double bonds. These hydrocarbons' electrons pz orbitals delocalize and form a delocalized bonding π orbital with a π* antibonding orbital. The delocalized π orbital is the highest occupied molecular orbital (HOMO), and the π* orbital is the lowest unoccupied molecular orbital (LUMO). In organic semiconductor physics, the HOMO takes the role of the valence band while the LUMO serves as the conduction band. The energy separation between the HOMO and LUMO energy levels is considered the band gap of organic electronic materials and is typically in the range of 1–4 eV.

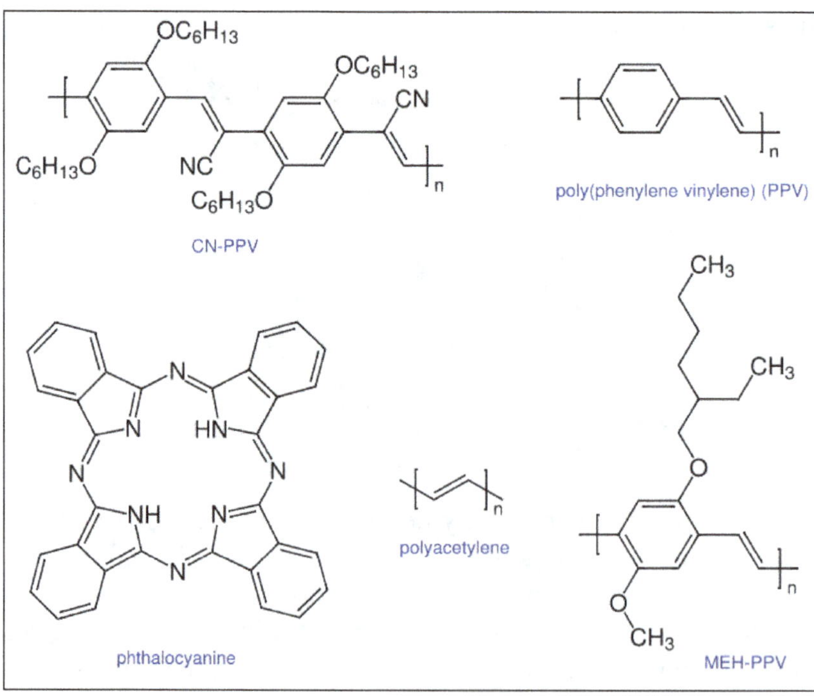


Examples of organic photovoltaic materials.

All light with energy greater than the band gap of the material can be absorbed, though there is a trade-off to reducing the band gap as photons absorbed with energies higher than the band gap will thermally give off its excess energy, resulting in lower voltages and power conversion efficiencies. When these materials absorb a photon, an excited state is created and confined to a molecule or a region of a polymer chain. The excited state can be regarded as an exciton, or an electron-hole pair bound together by electrostatic interactions. In photovoltaic cells, excitons are broken up into free electron-hole pairs by effective fields. The effective fields are set up by creating a heterojunction between two dissimilar materials. In organic photovoltaics, effective fields break up excitons by causing the electron to fall from the conduction band of the absorber to the conduction band of the acceptor molecule. It is necessary that the acceptor material has a conduction band edge that is lower than that of the absorber material.

Polymer solar cells usually consist of an electron- or hole-blocking layer on top of an indium tin oxide (ITO) conductive glass followed by electron donor and an electron acceptor (in the case of bulk heterojunction solar cells), a hole or electron blocking layer, and metal electrode on top. The nature and order of the blocking layers – as well as the nature of the metal electrode – depends on whether the cell follows a regular or an inverted device architecture. In an inverted cell, the electric charges exit the device in the opposite direction as in a normal device because the positive and negative electrodes are reversed. Inverted cells can utilize cathodes out of a more suitable material; inverted OPVs enjoy longer lifetimes than regularly structured OPVs, but they typically don't reach efficiencies as high as regular OPVs.

In bulk heterojunction polymer solar cells, light generates excitons. Subsequent charge separation in the interface between an electron donor and acceptor blend within the device's active layer. These charges then transport to the device's electrodes where the charges flow outside the cell, perform work and then re-enter the device on the opposite side. The cell's efficiency is limited by several factors, especially non-geminate recombination. Hole mobility leads to faster conduction across the active layer.

Organic photovoltaics are made of electron donor and electron acceptor materials rather than semiconductor p-n junctions. The molecules forming the electron donor region of organic PV cells, where exciton electron-hole pairs are generated, are generally conjugated polymers possessing delocalized $\pi$ electrons that result from carbon p orbital hybridization. These $\pi$ electrons can be excited by light in or near the visible part of the spectrum from the molecule's highest occupied molecular orbital (HOMO) to the lowest unoccupied molecular orbital (LUMO), denoted by a $\pi$ -$\pi^*$ transition. The energy bandgap between these orbitals determines which wavelength(s) of light can be absorbed.

Unlike in an inorganic crystalline PV cell material, with its band structure and delocalized electrons, excitons in organic photovoltaics are strongly bound with an energy between 0.1 and 1.4 eV. This strong binding occurs because electronic wave functions in organic molecules are more localized, and electrostatic attraction can thus keep the electron and hole together as an exciton. The electron and hole can be dissociated by providing an interface across which the chemical potential of electrons decreases. The material that absorbs the photon is the donor, and the material acquiring the electron is called the acceptor. The polymer chain is the donor and the fullerene is the acceptor. Even after dissociation, the electron and hole may still be joined as a "geminate pair", and an electric field is then required to separate them. The electron and hole must be collected at contacts. If charge carrier mobility is insufficient, the carriers will not reach the contacts, and instead recombine at trap sites or remain in the device as undesirable space charges that oppose the flow of new carriers. The latter problem can occur if electron and hole mobilities are not matched. In that case, space-charge limited photocurrent (SCLP) hampers device performance.

Organic photovoltaics can be fabricated with an active polymer and a fullerene-based electron acceptor. Illumination of this system by visible light leads to electron transfer from the polymer to a fullerene molecule. As a result, the formation of a photoinduced quasiparticle, or polaron ($P^+$), occurs on the polymer chain and the fullerene becomes a radical anion ($C_{60}^-$). Polarons are highly mobile and can diffuse away.

## Junction Types

The simplest organic PV device features a planar heterojunction. A film of organic active material (polymer or small molecule), of electron donor or electron acceptor type is sandwiched between contacts. Excitons created in the active material may diffuse before recombining and separate, hole and electron diffusing to its specific collecting electrode. Because charge carriers have diffusion lengths of just 3–10 nm in typical amorphous organic semiconductors, planar cells must be thin, but the thin cells absorb light less well. Bulk heterojunctions (BHJs) address this shortcoming. In a BHJ, a blend of electron donor and acceptor materials is cast as a mixture, which then phase-separates. Regions of each material in the device are separated by only several nanometers, a distance suited for carrier diffusion. BHJs require sensitive control over materials morphology on the nanoscale. Important variables include materials, solvents and the donor-acceptor weight ratio.

The next logical step beyond BHJs are ordered nanomaterials for solar cells, or ordered heterojunctions (OHJs). OHJs minimize the variability associated with BHJs. OHJs are generally hybrids of ordered inorganic materials and organic active regions. For example, a photovoltaic polymer can be deposited into pores in a ceramic such as $TiO_2$. Since holes still must diffuse the length of the pore through the polymer to a contact, OHJs suffer similar thickness limitations. Mitigating the hole mobility bottleneck is key to further enhancing device performance of OHJ's.

## Single Layer

Single layer organic photovoltaic cells are the simplest form. These cells are made by sandwiching a layer of organic electronic materials between two metallic conductors, typically a layer of indium tin oxide (ITO) with high work function and a layer of low work function metal such as Aluminum, Magnesium or Calcium.

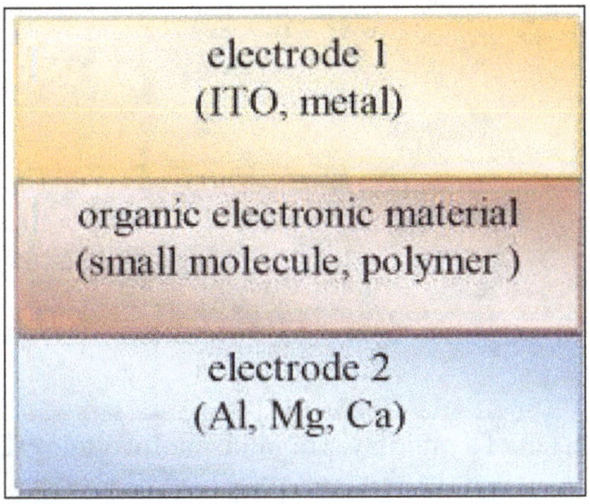


Sketch of a single layer organic photovoltaic cell.

The difference of work function between the two conductors sets up an electric field in the organic layer. When the organic layer absorbs light, electrons will be excited to the LUMO and leave holes in the HOMO, thereby forming excitons. The potential created by the different work functions helps to split the exciton pairs, pulling electrons to the positive electrode (an electrical conductor used to make contact with a non-metallic part of a circuit) and holes to the negative electrode.

Examples:

In 1958 the photovoltaic effect or the creation of voltage of a cell based on magnesium phthalocyanine (MgPc)—a macrocyclic compound having an alternating nitrogen atom-carbon atom ring structure—was discovered to have a photovoltage of 200 mV. An Al/MgPc/Ag cell obtained photovoltaic efficiency of 0.01% under illumination at 690 nm.

Conjugated polymers were also used in this type of photovoltaic cell. One device used polyacetylene as the organic layer, with Al and graphite, producing an open circuit voltage of 0.3 V and a charge collection efficiency of 0.3%. An Al/poly(3-nethyl-thiophene)/Pt cell had an external quantum yield of 0.17%, an open circuit voltage of 0.4 V and a fill factor of 0.3. An ITO/PPV/Al cell showed an open circuit voltage of 1 V and a power conversion efficiency of 0.1% under white-light illumination.

## Issues of Single Layer

Single layer organic solar cells do not work well. They have low quantum efficiencies (<1%) and low power conversion efficiencies (<0.1%). A major problem with them is that the electric field resulting from the difference between the two conductive electrodes is seldom sufficient to split the excitons. Often the electrons recombine with the holes without reaching the electrode.

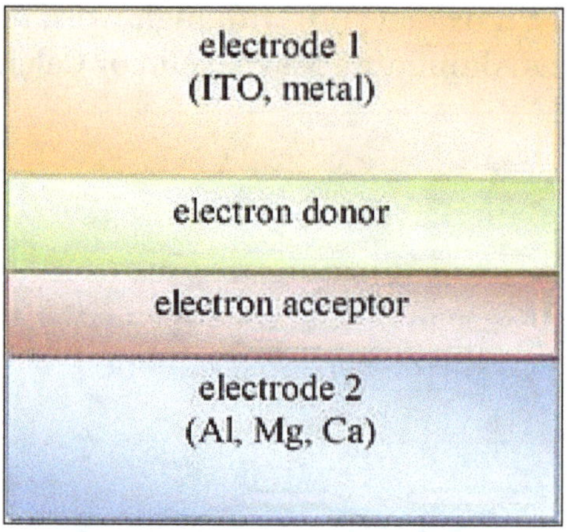


Sketch of a multilayer organic photovoltaic cell.

## Bilayer

Bilayer cells contain two layers in between the conductive electrodes. The two layers have different electron affinity and ionization energies, therefore electrostatic forces are generated at the interface between the two layers. Light must create excitons in this small charged region for an efficient charge separation and collecting. The materials are chosen to make the differences large enough that these local electric fields are strong, which splits excitons much more efficiently than single layer photovoltaic cells. The layer with higher electron affinity and ionization potential is the electron acceptor, and the other layer is the electron donor. This structure is also called a planar donor-acceptor heterojunction.

Examples:

$C_{60}$ has high electron affinity, making it a good acceptor. A $C_{60}$/MEH-PPV double layer cell had a relatively high fill factor of 0.48 and a power conversion efficiency of 0.04% under monochromatic illumination. PPV/$C_{60}$ cells displayed a monochromatic external quantum efficiency of 9%, a power conversion efficiency of 1% and a fill factor of 0.48.

Perylene derivatives display high electron affinity and chemical stability. A layer of copper phthalocyanine (CuPc) as electron donor and perylene tetracarboxylic derivative as electron acceptor, fabricating a cell with a fill factor as high as 0.65 and a power conversion efficiency of 1% under simulated AM2 illumination. Halls et al. fabricated a cell with a layer of bis(phenethylimido) perylene over a layer of PPV as the electron donor. This cell had peak external quantum efficiency of 6% and power conversion efficiency of 1% under monochromatic illumination, and a fill factor of up to 0.6.

## Issues of Bilayer

The diffusion length of excitons in organic electronic materials is typically on the order of 10 nm. In order for most excitons to diffuse to the interface of layers and split into carriers, the layer thickness should be in the same range as the diffusion length. However, a polymer layer typically needs a thickness of at least 100 nm to absorb enough light. At such a large thickness, only a small fraction of the excitons can reach the heterojunction interface.

## Discrete Heterojunction

A three-layer (two acceptor and one donor) fullerene-free stack achieved a conversion efficiency of 8.4%. The implementation produced high open-circuit voltages and absorption in the visible spectra and high short-circuit currents. Quantum efficiency was above 75% between 400 nm and 720 nm wavelengths, with an open-circuit voltage around 1 V.

## Bulk Heterojunction

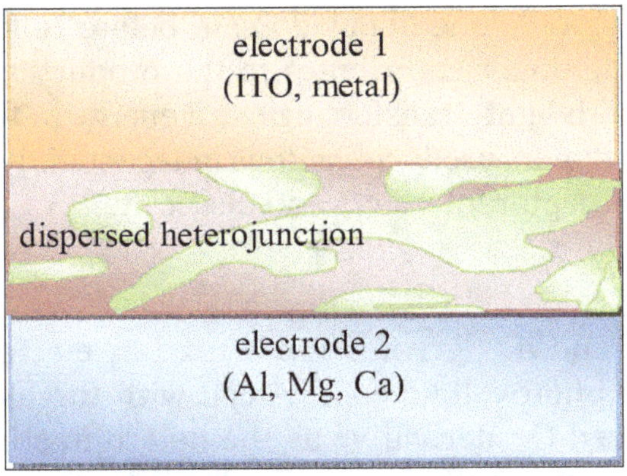


Sketch of a dispersed junction photovoltaic cell.

Bulk heterojunctions have an absorption layer consisting of a nanoscale blend of donor and acceptor materials. The domain sizes of this blend are on the order of nanometers, allowing for excitons with short lifetimes to reach an interface and dissociate due to the large donor-acceptor interfacial area. However, efficient bulk heterojunctions need to maintain large enough domain sizes to form a percolating network that allows the donor materials to reach the hole transporting electrode and the acceptor materials to reach the electron transporting electrode (Electrode 2). Without this percolating network, charges might be trapped in a donor or acceptor rich domain and undergo recombination. Bulk heterojunctions have an advantage over layered photoactive structures because they can be made thick enough for effective photon absorption without the difficult processing involved in orienting a layered structure while retaining similar level of performances.

Bulk heterojunctions are most commonly created by forming a solution containing the two components, casting (e.g. drop casting and spin coating) and then allowing the two phases to separate, usually with the assistance of an annealing step. The two components will self-assemble into an interpenetrating network connecting the two electrodes. They are normally composed of a conjugated molecule based donor and fullerene based acceptor. The nanostructural morphology of bulk heterojunctions tends to be difficult to control, but is critical to photovoltaic performance.

After the capture of a photon, electrons move to the acceptor domains, then are carried through the device and collected by one electrode, and holes move in the opposite direction and collected at the other side. If the dispersion of the two materials is too fine, it will result in poor charge transfer through the layer.

Most bulk heterojunction cells use two components, although three-component cells have been explored. The third component, a secondary p-type donor polymer, acts to absorb light in a different region of the solar spectrum. This in theory increases the amount of absorbed light. These ternary cells operate through one of three distinct mechanisms: charge transfer, energy transfer or parallel-linkage.

In charge transfer, both donors contribute directly to the generation of free charge carriers. Holes pass through only one donor domain before collection at the anode. In energy transfer, only one donor contributes to the production of holes. The second donor acts solely to absorb light, transferring extra energy to the first donor material. In parallel linkage, both donors produce excitons independently, which then migrate to their respective donor/acceptor interfaces and dissociate.

Examples:

Fullerenes such as $C_{60}$ and its derivatives are used as electron acceptor materials in bulk heterojunction photovoltaic cells. A cell with the blend of MEH-PPV and a methano-functionalized $C_{60}$ derivative as the heterojunction, ITO and Ca as the electrodes showed a quantum efficiency of 29% and a power conversion efficiency of 2.9% under monochromatic illumination. Replacing MEH-PPV with P3HT produced a quantum yield of 45% under a 10 V reverse bias. Further advances in modifying the electron acceptor has resulted in a device with a power conversion efficiency of 10.61% with a blend of $PC_{71}BM$ as the electron acceptor and PTB7-Th as the electron donor.

Polymer/polymer blends are also used in dispersed heterojunction photovoltaic cells. A blend of CN-PPV and MEH-PPV with Al and ITO as the electrodes, yielded peak monochromatic power conversion efficiency of 1% and fill factor of 0.38.

Dye sensitized photovoltaic cells can also be considered important examples of this type.

## Issues of Bulk Heterojunction

Fullerenes such as $PC_{71}BM$ are often the electron acceptor materials found in high performing bulk heterojunction solar cells. However, these electron acceptor materials very weakly absorb visible light, decreasing the volume fraction occupied by the strongly absorbing electron donor material. Furthermore, fullerenes have poor electronic tunability, resulting in restrictions placed on the development of conjugated systems with more appealing electronic structures for higher voltages. Recent research has been done on trying to replace these fullerenes with organic molecules that can be electronically tuned and contribute to light absorption.

## Graded Heterojunction

The electron donor and acceptor are mixed in such a way that the gradient is gradual. This architecture combines the short electron travel distance in the dispersed heterojunction with the advantage of the charge gradient of the bilayer technology.

Examples:

A cell with a blend of CuPc and $C_{60}$ showed a quantum efficiency of 50% and a power conversion efficiency of 2.1% using 100 mW/cm$^2$ simulated AM1.5G solar illumination for a graded heterojunction.

## Continuous Junction

Similar to the graded heterojunction the continuous junction concept aims at realizing a gradual transition from an electron donor to an electron acceptor. However, the acceptor material is prepared directly from the donor polymer in a post-polymerization modification step.

## Production

Since its active layer largely determines device efficiency, this component's morphology received much attention.

If one material is more soluble in the solvent than the other, it will deposit first on top of the substrate, causing a concentration gradient through the film. This has been demonstrated for poly-3-hexyl thiophene (P3HT), phenyl-$C_{61}$-butyric acid methyl ester (PCBM) devices where the PCBM tends to accumulate towards the device's bottom upon spin coating from ODCB solutions. This effect is seen because the more soluble component tends to migrate towards the "solvent rich" phase during the coating procedure, accumulating the more soluble component towards the film's bottom, where the solvent remains longer. The thickness of the generated film affects the phases segregation because the dynamics of crystallization and precipitation are different for more concentrated solutions or faster evaporation rates (needed to build thicker devices). Crystalline P3HT enrichment closer to the hole-collecting electrode can only be achieved for relatively thin (100 nm) P3HT/PCBM layers.

The gradients in the initial morphology are then mainly generated by the solvent evaporation rate and the differences in solubility between the donor and acceptor inside the blend. This dependence on solubility has been clearly demonstrated using fullerene derivatives and P3HT. When using solvents which evaporate at a slower rate (as chlorobenzene (CB) or dichlorobenzene (DCB)) you can get larger degrees of vertical separation or aggregation while solvents that evaporate quicker produce a much less effective vertical separation. Larger solubility gradients should lead to more effective vertical separation while smaller gradients should lead to more homogeneous films. These two effects were verified on P3HT:PCBM solar cells.

The solvent evaporation speed as well as posterior solvent vapor or thermal annealing procedures were also studied. Blends such as P3HT:PCBM seem to benefit from thermal annealing procedures, while others, such as PTB7:PCBM, seem to show no benefit. In P3HT the benefit seems to come from an increase of crystallinity of the P3HT phase which is generated through an expulsion of PCBM molecules from within these domains. This has been demonstrated through studies of PCBM miscibility in P3HT as well as domain composition changes as a function of annealing times.

The above hypothesis based on miscibility does not fully explain the efficiency of the devices as solely pure amorphous phases of either donor or acceptor materials never exist within bulk heterojunction devices. A 2010 paper suggested that current models that assume pure phases and discrete interfaces might fail given the absence of pure amorphous regions. Since current models assume phase separation at interfaces without any consideration for phase purity, the models might need to be changed.

The thermal annealing procedure varies depending on precisely when it is applied. Since vertical species migration is partly determined by the surface tension between the active layer and either air or another layer, annealing before or after the deposition of additional layers (most often the metal cathode) affects the result. In the case of P3HT:PCBM solar cells vertical migration is improved when cells are annealed after the deposition of the metal cathode.

Donor or acceptor accumulation next to the adjacent layers might be beneficial as these accumulations can lead to hole or electron blocking effects which might benefit device performance. In 2009 the difference in vertical distribution on P3HT:PCBM solar cells was shown to cause problems with electron mobility which ends up with the yielding of very poor device efficiencies. Simple changes to device architecture – spin coating a thin layer of PCBM on top of the P3HT – greatly enhance cell reproducibility, by providing reproducible vertical separation between device components. Since higher contact between the PCBM and the cathode is required for better efficiencies, this largely increases device reproducibility.

According to neutron scattering analysis, P3HT:PCBM blends have been described as "rivers" (P3HT regions) interrupted by "streams" (PCBM regions).

## Solvent Effects

Conditions for spin coating and evaporation affect device efficiency. Solvent and additives influence donor-acceptor morphology. Additives slow down evaporation, leading to more crystalline polymers and thus improved hole conductivities and efficiencies. Typical additives include 1,8-octanedithiol, ortho-dichlorobenzene, 1,8-diiodooctane (DIO), and nitrobenzene. The DIO effect was attributed to the selective solubilization of PCBM components, modifies fundamentally the average hopping distance of electrons, and thus improves electron mobility. Additives can also lead to big increases in efficiency for polymers. For HXS-1/PCBM solar cells, the effect was correlated with charge generation, transport and shelf-stability. Other polymers such as PTTBO also benefit significantly from DIO, achieving PCE values of more than 5% from around 3.7% without the additive.

Polymer Solar Cells fabricated from chloronaphthalene (CN) as a co-solvent enjoy a higher efficiency than those fabricated from the more conventional pure chlorobenzene solution. This is because the donor-acceptor morphology changes, which reduces the phase separation between donor polymer and fullerene. As a result, this translates into high hole mobilities. Without co-solvents, large domains of fullerene form, decreasing photovoltaic performance of the cell due to polymer aggregation in solution. This morphology originates from the liquid-liquid phase separation during drying; solve evaporation causes the mixture to enter into the spinodal region, in which there are significant thermal fluctuations. Large domains prevent electrons from being collected efficiently (decreasing PCE).

Small differences in polymer structure can also lead to significant changes in crystal packing that inevitably affect device morphology. PCPDTBT differs from PSBTBT caused by the difference in bridging atom between the two polymers (C vs. Si), which implies that better morphologies are achievable with PCPDTBT:PCBM solar cells containing additives as opposed to the Si system which achieves good morphologies without help from additional substances.

## Self-assembled Cells

Supramolecular chemistry was investigated, using donor and acceptor molecules that assemble upon spin casting and heating. Most supramolecular assemblies employ small molecules. Donor and acceptor domains in a tubular structure appear ideal for organic solar cells.

Diblock polymers containing fullerene yield stable organic solar cells upon thermal annealing. Solar cells with pre-designed morphologies resulted when appropriate supramolecular interactions are introduced.

Progress on BCPs containing polythiophene derivatives yield solar cells that assemble into well defined networks. This system exhibits a PCE of 2.04%. Hydrogen bonding guides the morphology.

Device efficiency based on co-polymer approaches have yet to cross the 2% barrier, whereas bulk-heterojunction devices exhibit efficiencies >7% in single junction configurations.

Fullerene-grafted rod-coil block copolymers have been used to study domain organization.

Supramolecular approaches to organic solar cells provide understanding about the macromolecular forces that drive domain separation.

## Transparent Polymer Cells

Transparent or semi-transparent PSCs allow for the absorption of low- or high-energy photons outside the visible spectrum, thus optimizing its sunlight harnessing capabilities and covering a broader absorption spectra. These types of PSCs are ideal for capturing near-infrared or ultraviolet photons due to its low inherent sensitivity to photons within the visible spectrum. Typical PSCs utilize opaque metal electrodes that limit its transparency, and thus its performance. The absorber layer of PSCs are intrinsically semi-transparent. Thus, one approach to achieving a visibly transparent PSC is to modify the top electrode to make it more transparent. Materials such as ITO, ultra-thin metals, metal grids, graphene, and carbon nanotubes have been used to fabricate semi-transparent top electrodes. Yet, the performance of transparent PSCs have shown to be lacking when compared to their opaque electrode PSC counterparts. When the top electrode is made transparent, the cell's ability to trap the electromagnetic field in the absorber layer decreases, resulting in a low PCE. An extensive amount of research is currently being conducted to improve the PCE of such cells. These types of PSCs have been applied to building-integrated photovoltaics, tandem devices, and portable electronics.

## Infrared Polymer Cells

Infrared cells preferentially absorb light in the infrared range rather than visible wavelengths. A study developed infrared-transparent PSCs with a CNT film top electrode on the back side and an ITO glass layer on the front side allowing for optical transmittance from both sides of the cell. A ZnO layer was placed on top of the ITO with a P3HT:PCBM layer being added to the ZnO, thus creating an ITO/ZnO/P3HT:PCBM/CNT (bottom to top) cell. It was observed that the top CNT electrode and bottom ITO electrode both exhibited 80% transmittance within a 500 nm to 2.5 um spectra. The cell itself had an optical transmittance of 80% in the 670 nm to 1.2 um range, 60% in the 1.2 um to 2.5 um range. Conversely, a control cell with an Ag top electrode resulted in no transmittance within this spectra. Additionally, the cell had a relatively transmittance in the visible region due to the high visible absorbance of the P3HT:PCBM layer. Such cells can be applied to tandem devices and the vertical assembly of PSCs.

As of 2012, infrared cells were nearly 70% transparent to visible light. The cells allegedly can be made in high volume at low cost using solution processing. The cells employ silver nanowire/titanium dioxide composite films as the top electrode, replacing conventional opaque metal electrodes. With this combination, 4% power-conversion efficiency was achieved.

In 2014, near-infrared polymer solar cells based on a copolymer of naphthodithiophene diimide and bithiophene (PNDTI-BT-DT) were fabricated in combination with PTB7 as an electron donor. Both PNDTI-BT-DT and PTB7 formed a crystalline structure in the blend films similar to in the pristine films, leading to the efficient charge generation contributed from both polymers.

Much research has been focused on developing a transparent top electrode for PSCs. However, a 2017 study explored optimizing the active layer of semi-transparent PSCs. The researchers proposed a semi-transparent PSC with enhanced efficiency that utilizes both narrow bandgap polymer donor, PTB7-Th, and non-fullerene acceptor, IHIC. The results of this study showed that the proposed PSC exhibited high transmittance and absorption in the infrared spectrum but low absorption in the visible spectrum. This cell showed to be relatively stable and have a maximum PCE of 9.77%, which, as of 2017, is the highest reported PCE value.

## Typical Current-voltage Behavior and Power Conversion Efficiency

Organic photovoltaics, similar to inorganic photovoltaics, are generally characterized through current-voltage analysis. This analysis provides multiple device metrics values that are used to understand device performance. One of the most crucial metrics is the Power Conversion Efficiency (PCE).

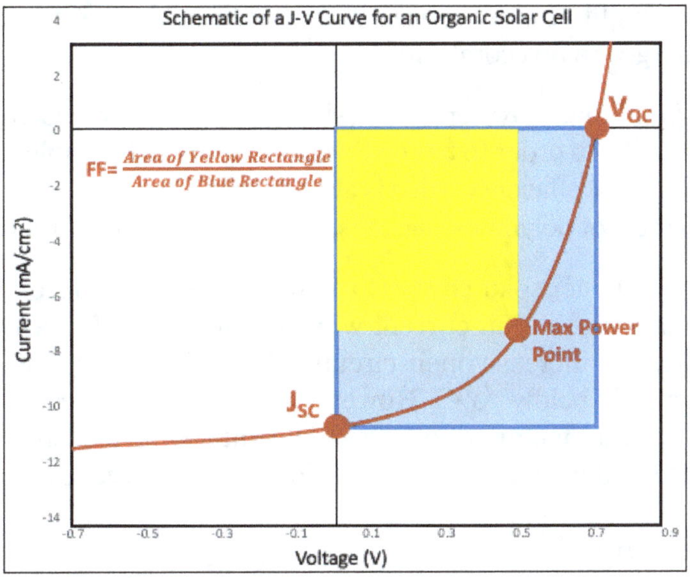


PCE (η) is proportional to the product of the short-circuit current ($J_{SC}$), the open circuit voltage ($V_{OC}$), and the fill factor (FF), all of which can be determined from a current-voltage curve.

$$\eta = \frac{V_{OC} \times J_{SC} \times FF}{P_{in}}$$

Where $P_{in}$ is the incident solar power.

The short circuit current ($J_{sc}$), is the maximum photocurrent generation value. It corresponds to the y-intercept value of standard current-voltage curve in which current is plotted along the y-axis and voltage is plotted along the x-axis. Within organic solar cells, the short circuit current can be impacted by a variety of material factors. These include the mobility of charge carriers, the optical absorption profile and general energetic driving forces that lead to a more efficient extraction of charge carriers

The open circuit voltage ($V_{oc}$) is the voltage when there is no current running through the device. This corresponds to the x-intercept on a current-voltage curve. Within bulk heterojunction organic photovoltaic devices, this value is highly dependent on HOMO and LUMO energy levels and work functions for the active layer materials

Since power is the product of voltage and current, the maximum power point occurs when the product between voltage and current is maximized.

The fill factor, FF, can be thought of as the "squareness" of a current voltage curve. It is the quotient of the maximum power value and the product of the open circuit voltage and short circuit current. This is shown in the image above as the ratio of the area of the yellow rectangle to the greater blue rectangle. For organic photovoltaics, this fill factor is essentially a measure of how efficiently generated charges are extracted from the device. This can be thought of as a "competition" between charges transporting through the device, and charges that recombine.

A major issue surrounding polymer solar cells is the low Power Conversion Efficiency (PCE) of fabricated cells. In order to be considered commercially viable, PSCs must be able to achieve at least 10–15% efficiency—this is already much lower than inorganic PVs. However, due to the low cost of polymer solar cells, a 10–15% efficiency is commercially viable.

Recent advances in polymer solar cell performance have resulted from compressing the bandgap to enhance short-circuit current while lowering the Highest Occupied Molecular Orbital (HOMO) to increase open-circuit voltage. However, PSCs still suffer from low fill factors (typically below 70%). However, as of 2013, researchers have been able to fabricate PSCs with fill factors of over 75%. Scientists have been able to accomplish via an inverted BHJ and by using nonconventional donor/acceptor combinations.

## Commercialization

Polymer solar cells have yet to commercially compete with silicon solar cells and other Thin film cells. The present efficiency of polymer solar cells lies near 10%, well below silicon cells. Polymer solar cells also suffer from environmental degradation, lacking effective protective coatings.

Further improvements in performance are needed to promote charge carrier diffusion; transport must be enhanced through control of order and morphology; and interface engineering must be applied to the problem of charge transfer across interfaces.

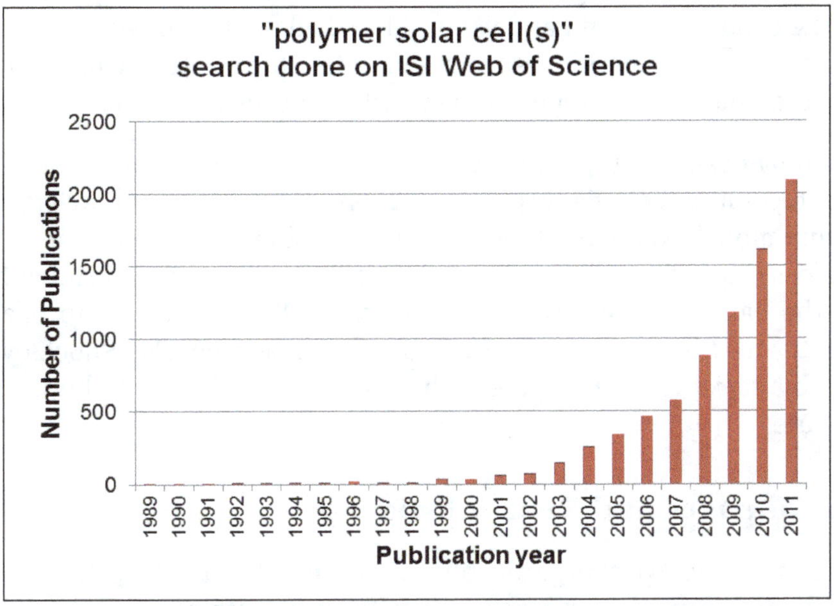


Number of scientific publications contributing to the subject "polymer solar cell(s)" by year. Search done through ISI, Web of Science.

Research is being conducted into using tandem architecture in order to increase efficiency of polymer solar cells. Similar to inorganic tandem architecture, organic tandem architecture is expected to increase efficiency. Compared with a single-junction device using low-bandgap materials, the tandem structure can reduce heat loss during photon-to-electron conversion.

Polymer solar cells are not widely produced commercially. Starting in 2008, Konarka Technologies started production of polymer-fullerene solar cells. The initial modules were 3–5% efficient, and only last for a few years. Konarka has since filed for bankruptcy, as those polymer solar cells were unable to penetrate the PV market.

PSCs also still suffer from low fill factors (typically below 70%). However, as of 2013, researchers have been able to fabricate PSCs with fill factors of over 75%. Scientists have been able to accomplish via an inverted BHJ and by using nonconventional donor/acceptor combinations.

However, efforts are being made to upscale manufacturing of polymer solar cells, in order to decrease costs and also advocate for a practical approach for PSC production. Such efforts include full roll-to-roll solution processing. However, roll-to-roll solution processing is ill-suited for on-grid electricity production due to the short lifetime of polymer solar cells. Therefore, commercial applications for polymer solar cells still include primarily consumer electronics and home appliances.

## Modeling Organic Solar Cells

Organic semiconductors are highly disordered materials with no long range order. This means that the conduction band and valance band edges are not well defined. Furthermore, this physical and energetic disorder generates trap states in which photogenerated electrons and holes can become trapped and then eventually recombine.

Key to accurately describing organic solar cells in a device model is to include carrier trapping and recombination via trap states. A commonly used approach is to use an effective medium model, where by standard drift diffusion equations are used to describe transport across the device. Then, an exponential tail of trap states is introduced which decays into the band gap from the mobility edges. To describe capture/escape from these trap states the Shockley–Read–Hall (SRH) can be used. The Shockley-Read-Hall mechanism able to reproduce polymer:fullerene device behavior in both time domain and steady state.

## Current Challenges and Recent Progress

Difficulties associated with organic photovoltaic cells include their low external quantum efficiency (up to 70%) compared to inorganic photovoltaic devices, despite having good internal quantum efficiency; this is due to insufficient absorption with active layers on the order of 100 nanometers. Instabilities against oxidation and reduction, recrystallization and temperature variations can also lead to device degradation and decreased performance over time. This occurs to different extents for devices with different compositions, and is an area into which active research is taking place.

Other important factors include the exciton diffusion length, charge separation and charge collection which are affected by the presence of impurities.

## Charge Carrier Mobility and Transport

Especially for bulk heterojunction solar cells, understanding charge carrier transport is vital in improving the efficiencies of organic photovoltaics. Currently, bulk heterojunction devices have imbalanced charge-carrier mobility, with the hole mobility being at least an order of magnitude lower than that of the electron mobility; this results in space charge build-up and a decrease in the fill factor and power conversion efficiency of a device. Due to having low mobility, efficient bulk heterojunction photovoltaics have to be designed with thin active layers to avoid recombination of the charge carriers, which is detrimental to absorption and scalability in processing. Simulations have demonstrated that in order to have a bulk heterojunction solar cell with a fill factor above 0.8 and external quantum efficiency above 90%, there needs to be balanced charge carrier mobility to reduce a space charge effect, as well as an increase in charge carrier mobility a decrease in the bimolecular recombination rate constant.

## Effect of Film Morphology

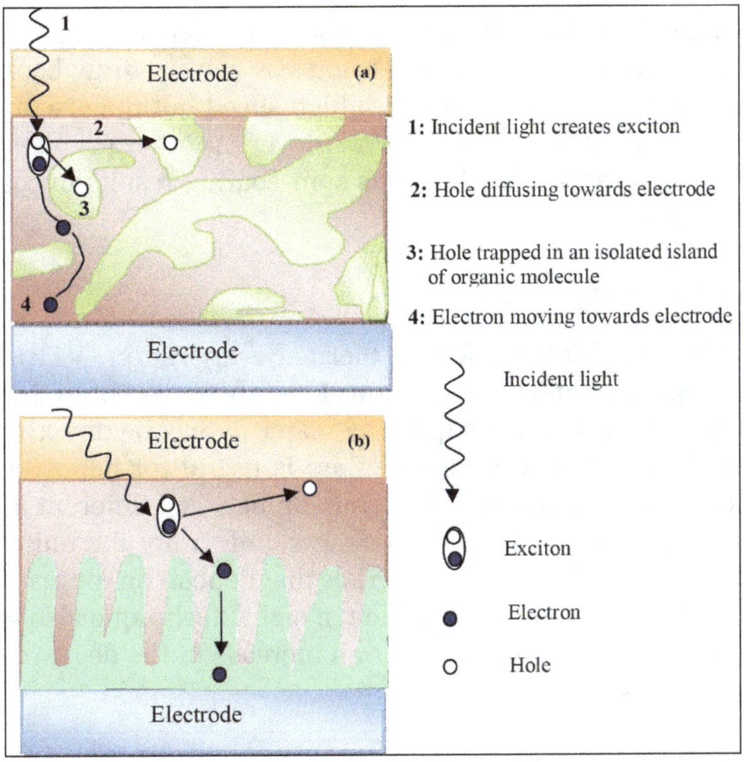


Highly folded heterojunction (a); heterojunction with controlled growth (b).

Dispersed heterojunctions of donor-acceptor organic materials have high quantum efficiencies compared to the planar hetero-junction, because in dispersed heterojunctions it is more likely for an exciton to find an interface within its diffusion length. Film morphology can also have a drastic effect on the quantum efficiency of the device. Rough surfaces and the presence of voids can increase the series resistance and also the chance of short-circuiting. Film morphology and, as a result, quantum efficiency can be improved by annealing of a device after covering it by a ~1000 Å thick metal cathode. Metal film on top of the organic film applies stresses on the organic film, which helps to prevent the morphological relaxation in the organic film. This gives more densely packed films and at the same time allows the formation of phase-separated interpenetrating donor-acceptor interface inside the bulk of organic thin film.

## Controlled Growth Heterojunction

Charge separation occurs at the donor-acceptor interface. Whilst traveling to the electrode, a charge can become trapped recombine in a disordered interpenetrating organic material, resulting in decreased device efficiency. Controlled growth of the heterojunction provides better control over positions of the donor-acceptor materials, resulting in much greater power efficiency (ratio of output power to input power) than that of planar and highly disoriented hetero-junctions. Thus, the choice of suitable processing parameters in order to better control the structure and film morphology is highly desirable.

## Progress in Growth Techniques

Mostly organic films for photovoltaic applications are deposited by spin coating and vapor-phase deposition. However each method has certain draw backs, spin coating technique can coat larger surface areas with high speed but the use of solvent for one layer can degrade the already existing polymer layer. Another problem is related with the patterning of the substrate for device as spin-coating results in coating the entire substrate with a single material.

## Vacuum Thermal Evaporation

Another deposition technique is vacuum thermal evaporation (VTE) which involves the heating of an organic material in vacuum. The substrate is placed several centimeters away from the source so that evaporated material may be directly deposited onto the substrate, as shown in figure. This method is useful for depositing many layers of different materials without chemical interaction between different layers. However, there are sometimes problems with film-thickness uniformity and uniform doping over large-area substrates. In addition, the materials that deposit on the wall of the chamber can contaminate later depositions. This “line of sight” technique also can create holes in the film due to shadowing, which causes an increase in the device series-resistance and short circuit.

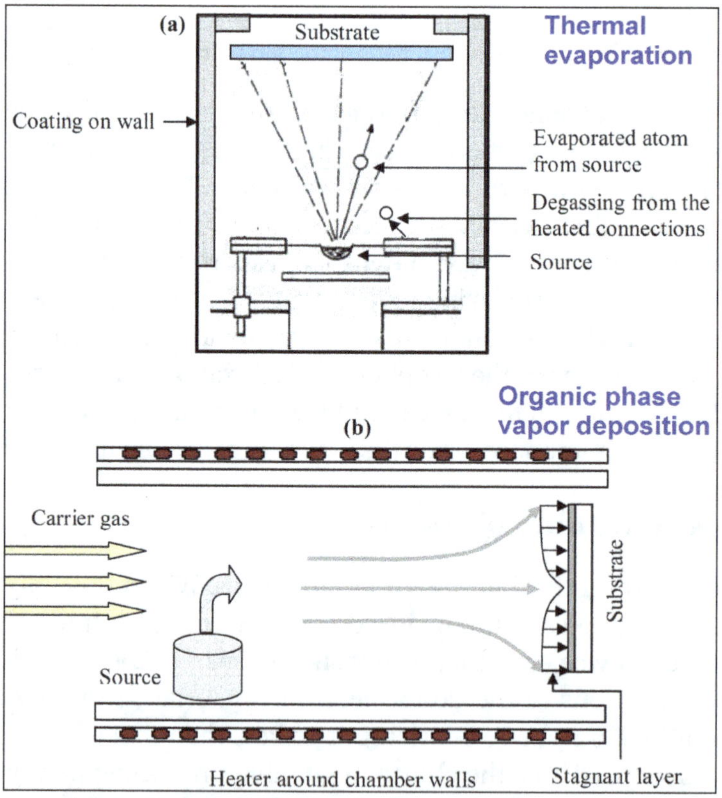


Vacuum thermal evaporation (a) and organic phase vapor deposition (b).

## Organic Vapor Phase Deposition

Organic vapor phase deposition allows better control of the structure and morphology of the film than vacuum thermal evaporation. The process involves evaporation of the organic material over a substrate in the presence of an inert carrier gas. The resulting film morphology can be tuned by changing the gas flow rate and the source temperature. Uniform films can be grown by reducing the carrier gas pressure, which will increase the velocity and mean free path of the gas, and as a result boundary layer thickness decreases. Cells produced by OVPD do not have issues related with contaminations from the flakes coming out of the walls of the chamber, as the walls are warm and do not allow molecules to stick to and produce a film upon them.

Another advantage over VTE is the uniformity in evaporation rate. This occurs because the carrier gas becomes saturated with the vapors of the organic material coming out of the source and then moves towards the cooled substrate, figure. Depending on the growth parameters (temperature of the source, base pressure and flux of the carrier gas) the deposited film can be crystalline or amorphous in nature. Devices fabricated using OVPD show a higher short-circuit current density than that of devices made using VTE. An extra layer of donor-acceptor hetero-junction at the top of the cell may block excitons, whilst allowing conduction of electron; resulting in improved cell efficiency.

## Organic Solar Ink

Organic solar ink is able to deliver higher performance in fluorescent lighting conditions in comparison to amorphous silicon solar cells, and said to have a 30% to 40% increase in indoor power density in comparison to the standard organic solar technology.

## Light Trapping

Various type of components are applied to increase light trapping (Light in-coupling) effects in thin organic solar cells. In addition to the flexibility of organic solar cells, by using flexible electrodes and substrates instead of ITO and glass respectively, fully flexible organic solar cells can be produced. By these use of flexible substrates and substrates, easier methods to provide light trapping effects to OPVs are introduced such as polymer electrodes with embedded scattering particles, nano imprinted polymer electrodes, patterned PET substrates and even optical display film commercialized for liquid crystal displays (LCD) as substrates. Much research will be taken for enhancing the performance of OPVs with the merit of easy light trapping structures processing.

## Use in Tandem Photovoltaics

Recent research and study has been done in utilizing an organic solar cell as the top cell in a hybrid tandem solar cell stack. Because organic solar cells have a higher band gap than traditional inorganic photovoltaics like silicon or CIGS, they can absorb higher energy photons without losing much of the energy due to thermalization, and thus operate at a higher voltage. The lower energy photons and higher energy photons that are unabsorbed pass through the top organic solar cell and are then absorbed by the bottom inorganic cell. Organic solar cells are also solution processible at low temperatures with a low cost of 10 dollars per square meter, resulting in a printable top cell that improves the overall efficiencies of existing, inorganic solar cell technologies. Much research has been done to enable the formation of such a hybrid tandem solar cell stack, including research in the deposition of semi-transparent electrodes that maintain low contact resistance while having high transparency.

### Recent Directions for Bulk Heterojunction Materials Research

One major area of current research is the use of non-fullerene acceptors. While fullerene acceptors have been the standard for most organic photovoltaics due to their compatibility within bulk heterojunction cell designs as well as their good transport properties, they do have some fallbacks that are leading researchers to attempt to find alternatives. Some negatives of fullerene acceptors include their instability, that they are somewhat limited in energy-tunability and they have poor optical absorption. Researchers have developed small molecule acceptors that due to their good energy tunability, can exhibit high open circuit voltages. However, there are still major challenges with non-fullerene acceptors, including the low charge carrier mobilities of small molecule acceptors, and that the sheer number of possible molecules is overwhelming for the research community.

Small molecules are also being heavily researched to act as donor materials, potentially replacing polymeric donors. Since small molecules do not vary in molecular weights the way polymers do, they would require less purification steps and are less susceptible to macromolecule defects and kinks that can create trap states leading to recombination. Recent research has shown that high-performing small molecular donor structures tend to have planar 2-D structures and can aggregate or self assemble. Sine performance of these devices is highly depended on active layer morphology, present research is continuing to investigate small molecule possibilities, and optimize device morphology through processes such as annealing for various materials.

## Solar Panel

Photovoltaic solar panels absorb sunlight as a source of energy to generate direct current electricity. A photovoltaic (PV) module is a packaged, connected assembly of photovoltaic solar cells available in different voltages and wattages. Photovoltaic modules constitute the photovoltaic array of a photovoltaic system that generates and supplies solar electricity in commercial and residential applications.

Solar PV modules (top) and two solar hot water panels (bottom) mounted on rooftops.

The most common application of solar energy collection outside agriculture is solar water heating systems.

## Theory and Construction

Photovoltaic modules use light energy (photons) from the Sun to generate electricity through the photovoltaic effect. Most modules use wafer-based crystalline silicon cells or Thin film cells. The structural (load carrying) member of a module can be either the top layer or the back layer. Cells must be protected from mechanical damage and moisture. Most modules are rigid, but semi-flexible ones based on Thin film cells are also available. The cells are connected electrically in series, one to another to a desired voltage, and then in parallel to increase amperage. The wattage of the module is the mathematical product of the voltage and the amperage of the module.

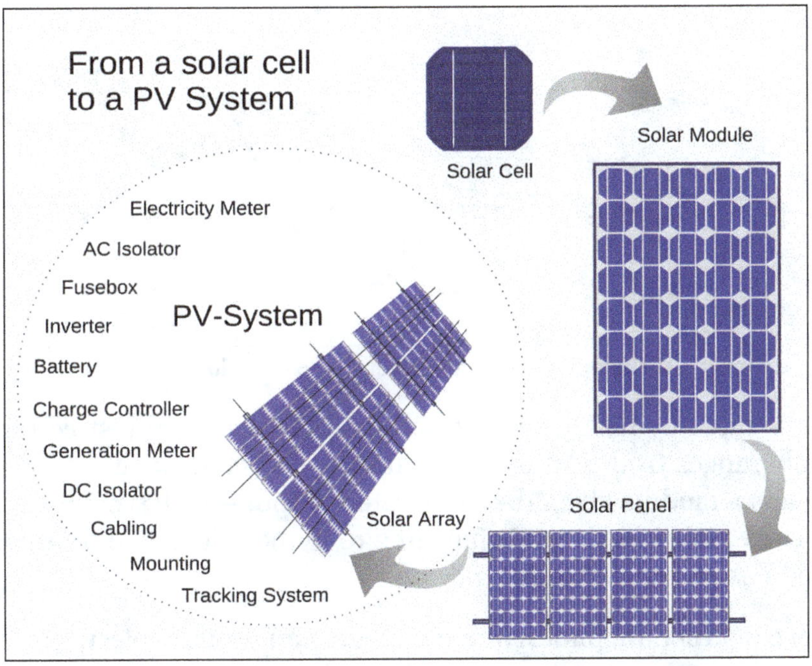


From a solar cell to a PV system.

A PV junction box is attached to the back of the solar panel and functions as its output interface. External connections for most photovoltaic modules use MC4 connectors to facilitate easy weatherproof connections to the rest of the system. A USB power interface can also be used.

Module electrical connections are made in series to achieve a desired output voltage or in parallel to provide a desired current capability (amperes) of the solar panel or the PV system. The conducting wires that take the current off the modules are sized according to the ampacity and may contain silver, copper or other non-magnetic conductive transition metals. Bypass diodes may be incorporated or used externally, in case of partial module shading, to maximize the output of module sections still illuminated.

Some special solar PV modules include concentrators in which light is focused by lenses or mirrors onto smaller cells. This enables the use of cells with a high cost per unit area (such as gallium arsenide) in a cost-effective way.

Solar panels also use metal frames consisting of racking components, brackets, reflector shapes, and troughs to better support the panel structure.

## Efficiencies

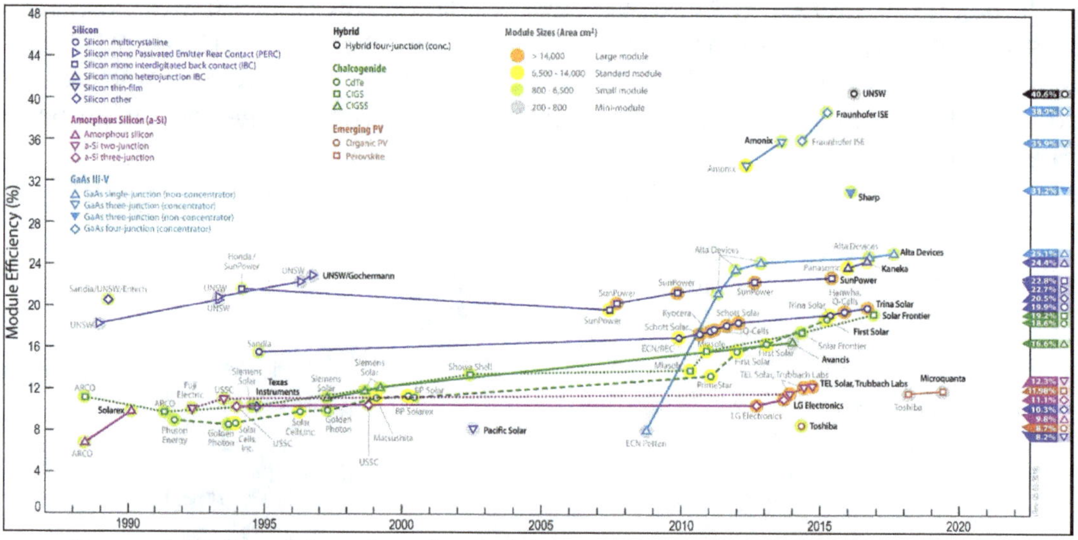


Reported timeline of champion solar module energy conversion efficiencies since 1988.

Each module is rated by its DC output power under standard test conditions (STC). Power typically ranges from 100 to 365 Watts (W). The efficiency of a module determines the area of a module given the same rated output – an 8% efficient 230 W module will have twice the area of a 16% efficient 230 W module. Some commercially available solar modules exceed 24% efficiency.

Depending on construction, photovoltaic modules can produce electricity from a range of frequencies of light, but usually cannot cover the entire solar range (specifically, ultraviolet, infrared and low or diffused light). Hence, much of the incident sunlight energy is wasted by solar modules, and they can give far higher efficiencies if illuminated with monochromatic light. Therefore, another design concept is to split the light into six to eight different wavelength ranges that will produce a different color of light, and direct the beams onto different cells tuned to those ranges. This has been projected to be capable of raising efficiency by 50%.

A single solar module can produce only a limited amount of power; most installations contain multiple modules adding voltages or current to the wiring and PV system. A photovoltaic system typically includes an array of photovoltaic modules, an inverter, a battery pack for energy storage, charge controller, interconnection wiring, circuit breakers, fuses, disconnect switches, voltage meters, and optionally a solar tracking mechanism. Equipment is carefully selected to optimize output, energy storage, reduce power loss during power transmission, and conversion from direct current to alternating current.

Scientists from Spectrolab, a subsidiary of Boeing, have reported development of multi-junction solar cells with an efficiency of more than 40%, a new world record for solar photovoltaic cells. The Spectrolab scientists also predict that concentrator solar cells could achieve efficiencies of more than 45% or even 50% in the future, with theoretical efficiencies being about 58% in cells with more than three junctions.

Currently, the best achieved sunlight conversion rate (solar module efficiency) is around 21.5% in new commercial products typically lower than the efficiencies of their cells in isolation. The most efficient mass-produced solar modules have power density values of up to 175 $W/m^2$ (16.22 $W/ft^2$).

Research by Imperial College, London has shown that solar panel efficiency is improved by studding the light-receiving semiconductor surface with aluminum nanocylinders, similar to the ridges on Lego blocks. The scattered light then travels along a longer path in the semiconductor, absorbing more photons to be converted into current. Although these nanocylinders have been used previously (aluminum was preceded by gold and silver), the light scattering occurred in the near infrared region and visible light was absorbed strongly. Aluminum was found to have absorbed the ultraviolet part of the spectrum, while the visible and near infrared parts of the spectrum were found to be scattered by the aluminum surface. This, the research argued, could bring down the cost significantly and improve the efficiency as aluminum is more abundant and less costly than gold and silver. The research also noted that the increase in current makes thinner film solar panels technically feasible without "compromising power conversion efficiencies, thus reducing material consumption".

- Solar panel efficiency can be calculated by MPP (maximum power point) value of solar panels.
- Solar inverters convert the DC power to AC power by performing the process of maximum power point tracking (MPPT): solar inverter samples the output

Power (I-V curve) from the solar cell and applies the proper resistance (load) to solar cells to obtain maximum power.

- MPP (Maximum power point) of the solar panel consists of MPP voltage (V mpp) and MPP current (I mpp): it is a capacity of the solar panel and the higher value can make higher MPP.

Micro-inverted solar panels are wired in parallel, which produces more output than normal panels wired in series, with the output of the series determined by the lowest performing panel. This is known as the "Christmas light effect". Micro-inverters work independently to enable each panel to contribute its maximum possible output for a given amount of sunlight.

## Technology

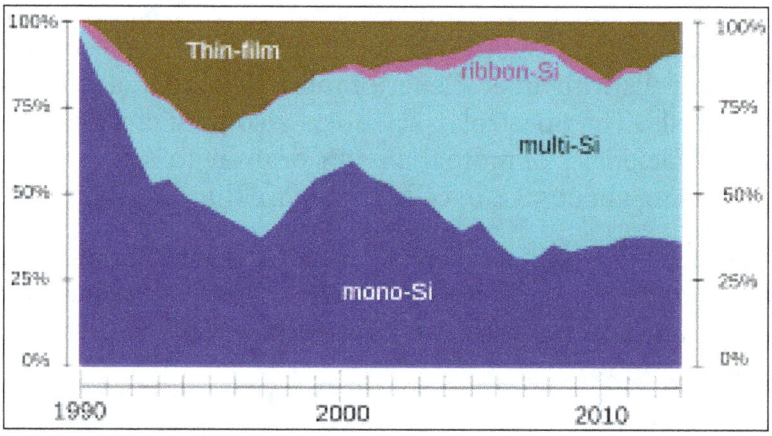


Market-share of PV technologies since 1990.

Most solar modules are currently produced from crystalline silicon (c-Si) solar cells made of multicrystalline and monocrystalline silicon. In 2013, crystalline silicon accounted for more than 90 percent of worldwide PV production, while the rest of the overall market is made up of Thin film technologies using cadmium telluride, CIGS and amorphous silicon.

Emerging, third generation solar technologies use advanced Thin film cells. They produce a relatively high-efficiency conversion for the low cost compared to other solar technologies. Also, high-cost, high-efficiency, and close-packed rectangular multi-junction (MJ) cells are preferably used in solar panels on spacecraft, as they offer the highest ratio of generated power per kilogram lifted into space. MJ-cells are compound semiconductors and made of gallium arsenide (GaAs) and other semiconductor materials. Another emerging PV technology using MJ-cells is concentrator photovoltaics ( CPV ).

## Thin Film

In rigid Thin film modules, the cell and the module are manufactured in the same production line. The cell is created on a glass substrate or superstrate, and the electrical connections are created in situ, a so-called "monolithic integration". The substrate or superstrate is laminated with an encapsulant to a front or back sheet, usually another sheet of glass. The main cell technologies in this category are CdTe, or a-Si, or a-Si+uc-Si tandem, or CIGS (or variant). Amorphous silicon has a sunlight conversion rate of 6–12%

Flexible thin film cells and modules are created on the same production line by depositing the photoactive layer and other necessary layers on a flexible substrate. If the substrate is an insulator (e.g. polyester or polyimide film) then monolithic integration can be used. If it is a conductor then another technique for electrical connection must be used. The cells are assembled into modules by laminating them to a transparent colourless fluoropolymer on the front side (typically ETFE or FEP) and a polymer suitable for bonding to the final substrate on the other side.

## Smart Solar Modules

Several companies have begun embedding electronics into PV modules. This enables performing MPPT for each module individually, and the measurement of performance data for monitoring and fault detection at module level. Some of these solutions make use of power optimizers, a DC-to-DC converter technology developed to maximize the power harvest from solar photovoltaic systems. As of about 2010, such electronics can also compensate for shading effects, wherein a shadow falling across a section of a module causes the electrical output of one or more strings of cells in the module to fall to zero, but not having the output of the entire module fall to zero.

## Performance and Degradation

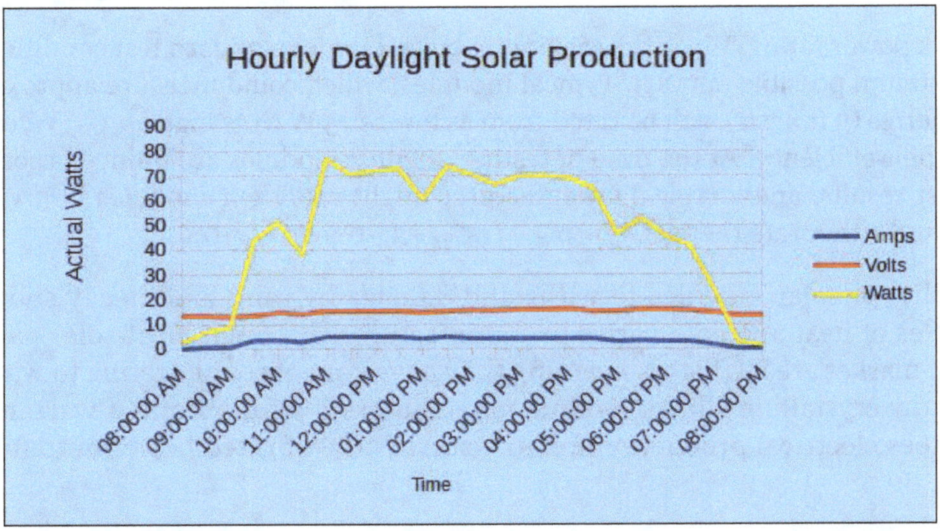


Module performance is generally rated under standard test conditions (STC): irradiance of 1,000 W/m², solar spectrum of AM 1.5 and module temperature at 25 °C. The actual voltage and current output of the module changes as lighting, temperature and load conditions change, so there is never one specific voltage, current, or wattage at which the module operates. Performance varies depending on time of day, amount of solar insolation, direction and tilt of modules, cloud cover, shading, state of charge, temperature, geographic location, and day of the year. The fluctuations in voltage and current can be logged with multimeter or data logger.

For optimum performance a solar panel needs to be made of similar modules oriented in the same direction perpendicular towards direct sunlight. The path of the sun varies by latitude and day of the year and can be studied using a sundial or a sunchart and tracked using a solar tracker. Differences in voltage or current of modules may affect the overall performance of a panel. Bypass diodes are used to circumvent broken or shaded panels to optimize output.

Electrical characteristics include nominal power ($P_{MAX}$, measured in W), open circuit voltage ($V_{OC}$), short circuit current ($I_{SC}$, measured in amperes), maximum power voltage ($V_{MPP}$), maximum power current ($I_{MPP}$), peak power, (watt-peak, $W_p$), and module efficiency (%).

Nominal voltage refers to the voltage of the battery that the module is best suited to charge; this is a leftover term from the days when solar modules were only used to charge batteries. Nominal voltage allows users, at a glance, to make sure the module is compatible with a given system.

Open circuit voltage or $V_{OC}$ is the maximum voltage that the module can produce when not connected to an electrical circuit or system. $V_{OC}$ can be measured with a voltmeter directly on an illuminated module's terminals or on its disconnected cable.

The peak power rating, $W_p$, is the maximum output under standard test conditions (not the maximum possible output). Typical modules, which could measure approximately 1 by 2 metres (3 ft × 7 ft), will be rated from as low as 75 W to as high as 350 W, depending on their efficiency. At the time of testing, the test modules are binned according to their test results, and a typical manufacturer might rate their modules in 5 W increments, and either rate them at +/- 3%, +/-5%, +3/-0% or +5/-0%.

The ability of solar modules to withstand damage by rain, hail, heavy snow load, and cycles of heat and cold varies by manufacturer, although most solar panels on the U.S. market are UL listed, meaning they have gone through testing to withstand hail. Many crystalline silicon module manufacturers offer a limited warranty that guarantees electrical production for 10 years at 90% of rated power output and 25 years at 80%.

Potential induced degradation (also called PID) is a potential induced performance degradation in crystalline photovoltaic modules, caused by so-called stray currents. This effect may cause power loss of up to 30%.

The largest challenge for photovoltaic technology is said to be the purchase price per watt of electricity produced. New materials and manufacturing techniques continue to improve the price to power performance. The problem resides in the enormous activation energy that must be overcome for a photon to excite an electron for harvesting purposes. Advancements in photovoltaic technologies have brought about the process of "doping" the silicon substrate to lower the activation energy thereby making the panel more efficient in converting photons to retrievable electrons.

Chemicals such as boron (p-type) are applied into the semiconductor crystal in order to create donor and acceptor energy levels substantially closer to the valence and conductor bands. In doing so, the addition of boron impurity allows the activation energy to decrease 20 fold from 1.12 eV to 0.05 eV. Since the potential difference ($E_B$) is so low, the boron is able to thermally ionize at room temperatures. This allows for free energy carriers in the conduction and valence bands thereby allowing greater conversion of photons to electrons.

## Maintenance

Solar panel conversion efficiency, typically in the 20% range, is reduced by dust, grime, pollen, and other particulates that accumulate on the solar panel. "A dirty solar panel can reduce its power capabilities by up to 30% in high dust/pollen or desert areas", says Seamus Curran, associate professor of physics at the University of Houston and director of the Institute for NanoEnergy, which specializes in the design, engineering, and assembly of nanostructures.

Paying to have solar panels cleaned is often not a good investment; researchers found panels that had not been cleaned, or rained on, for 145 days during a summer drought in California, lost only 7.4% of their efficiency. Overall, for a typical residential solar system of 5 kW, washing panels halfway through the summer would translate into a mere $20 gain in electricity production until the summer drought ends—in about 2 ½ months. For larger commercial rooftop systems, the financial losses are bigger but still rarely enough to warrant the cost of washing the panels. On average, panels lost a little less than 0.05% of their overall efficiency per day. There may also be occupational hazards of solar panel installation and maintenance.

## Recycling

Most parts of a solar module can be recycled including up to 95% of certain semiconductor materials or the glass as well as large amounts of ferrous and non-ferrous metals. Some private companies and non-profit organizations are currently engaged in take-back and recycling operations for end-of-life modules.

Recycling possibilities depend on the kind of technology used in the modules:

- Silicon based modules: Aluminum frames and junction boxes are dismantled manually at the beginning of the process. The module is then crushed in a mill and the different fractions are separated - glass, plastics and metals. It is possible to recover more than 80% of the incoming weight. This process can be performed by flat glass recyclers since morphology and composition of a PV module is similar to those flat glasses used in the building and automotive industry. The recovered glass for example is readily accepted by the glass foam and glass insulation industry.

- Non-silicon based modules: They require specific recycling technologies such as the use of chemical baths in order to separate the different semiconductor materials. For cadmium telluride modules, the recycling process begins by crushing the module and subsequently separating the different fractions. This recycling process is designed to recover up to 90% of the glass and 95% of the semiconductor materials contained. Some commercial-scale recycling facilities have been created in recent years by private companies. For aluminium flat plate reflector: the trendiness of the reflectors has been brought up by fabricating them using a thin layer (around 0.016 mm to 0.024 mm) of Aluminum coating present inside the non-recycled plastic food packages.

Since 2010, there is an annual European conference bringing together manufacturers, recyclers and researchers to look at the future of PV module recycling.

## Production

The production of PV systems has followed a classic learning curve effect, with significant cost reduction occurring alongside large rises in efficiency and production output.

| Top Module Producer | Shipments in 2014 (MW) |
|---|---|
| Yingli | 3,200 |
| Trina Solar | 2,580 |
| Sharp Solar | 2,100 |
| Canadian Solar | 1,894 |
| Jinko Solar | 1,765 |
| ReneSola | 1,728 |
| First Solar | 1,600 |
| Hanwha SolarOne | 1,280 |
| Kyocera | 1,200 |
| JA Solar | 1,173 |

In 2010, 15.9 GW of solar PV system installations were completed, with solar PV pricing survey and market research company PVinsights reporting growth of 117.8% in solar PV installation on a year-on-year basis.

With over 100% year-on-year growth in PV system installation, PV module makers dramatically increased their shipments of solar modules in 2010. They actively expanded their capacity and turned themselves into gigawatt GW players. According to

PVinsights, five of the top ten PV module companies in 2010 are GW players. Suntech, First Solar, Sharp, Yingli and Trina Solar are GW producers now, and most of them doubled their shipments in 2010.

The basis of producing solar panels revolves around the use of silicon cells. These silicon cells are typically 10-20% efficient at converting sunlight into electricity, with newer production models now exceeding 22%. In order for solar panels to become more efficient, researchers across the world have been trying to develop new technologies to make solar panels more effective at turning sunlight into energy.

In 2018, the world's top four solar module producers in terms of shipped capacity during the calendar year of 2018 were Jinko Solar, JA Solar, Trina Solar, Longi Solar, and Canadian Solar.

## Mounting and Tracking

Solar modules mounted on solar trackers.

Workers install residential rooftop solar panels.

Ground-mounted photovoltaic systems are usually large, utility-scale solar power plants. Their solar modules are held in place by racks or frames that are attached to ground-based mounting supports. Ground based mounting supports include:

- Pole mounts, which are driven directly into the ground or embedded in concrete.
- Foundation mounts, such as concrete slabs or poured footings.
- Ballasted footing mounts, such as concrete or steel bases that use weight to secure the solar module system in position and do not require ground penetration. This type of mounting system is well suited for sites where excavation is not possible such as capped landfills and simplifies decommissioning or relocation of solar module systems.

Roof-mounted solar power systems consist of solar modules held in place by racks or frames attached to roof-based mounting supports.

Roof-based mounting supports include:

- Rail mounts, which are attached directly to the roof structure and may use additional rails for attaching the module racking or frames.
- Ballasted footing mounts, such as concrete or steel bases that use weight to secure the panel system in position and do not require through penetration. This mounting method allows for decommissioning or relocation of solar panel systems with no adverse effect on the roof structure.
- All wiring connecting adjacent solar modules to the energy harvesting equipment must be installed according to local electrical codes and should be run in a conduit appropriate for the climate conditions.

Solar trackers increase the energy produced per module at the cost of mechanical complexity and increased need for maintenance. They sense the direction of the Sun and tilt or rotate the modules as needed for maximum exposure to the light. Alternatively, fixed racks hold modules stationary throughout the day at a given tilt (zenith angle) and facing a given direction (azimuth angle). Tilt angles equivalent to an installation's latitude are common. Some systems may also adjust the tilt angle based on the time of year. Similarly, to maximize total energy output, modules are often oriented to face south (in the Northern Hemisphere) or north (in the Southern Hemisphere). On the other hand, east- and west-facing arrays (covering an east-west facing roof, for example) may also be useful. Even though such installations might not produce the maximum possible total energy, their power output would likely be more consistent throughout the day and possibly larger during peak demand.

## Standards

Standards generally used in photovoltaic modules:

- IEC 61215 (crystalline silicon performance), 61646 (thin film performance) and 61730 (all modules, safety).
- ISO 9488 Solar energy—Vocabulary.
- UL 1703 from Underwriters Laboratories.
- UL 1741 from Underwriters Laboratories.
- UL 2703 from Underwriters Laboratories.
- CE mark.
- Electrical Safety Tester (EST) Series (EST-460, EST-22V, EST-22H, EST-110).

## Connectors

Outdoor solar panels usually include MC4 connectors. Automotive solar panels may also include a car lighter USB adapter. Indoor panels (including solar pv glasses, thin films and windows) can integrate microinverter (AC Solar panels).

## Applications

There are many practical applications for the use of solar panels or photovoltaics. It can first be used in agriculture as a power source for irrigation. In health care solar panels can be used to refrigerate medical supplies. It can also be used for infrastructure. PV modules are used in photovoltaic systems and include a large variety of electric devices:

- Photovoltaic power stations.
- Rooftop solar PV systems.
- Standalone PV systems.
- Solar hybrid power systems.
- Concentrated photovoltaics.
- Solar planes.
- Solar-powered water purification.
- Solar-pumped lasers.
- Solar vehicles.
- Solar panels on spacecrafts and space stations.

## Limitations

### Pollution and Energy in Production

Solar panel has been a well-known method of generating clean, emission free electricity. However, it produces only direct current electricity (DC), which is not what normal appliances use. Solar photovoltaic systems (solar PV systems) are often made of solar PV panels (modules) and inverter (changing DC to AC). Solar PV panels are mainly made of solar photovoltaic cells, which has no fundamental difference to the material for making computer chips. The process of producing solar PV cells (computer chips) is energy intensive and involves highly poisonous and environmental toxic chemicals. There are few solar PV manufacturing plants around the world producing PV modules with energy produced from PV. This measure greatly reduces the carbon footprint during the manufacturing process. Managing the chemicals used in the manufacturing process is subject to the factories' local laws and regulations.

### Impact on Electricity Network

With the increasing levels of rooftop photovoltaic systems, the energy flow becomes 2-way. When there is more local generation than consumption, electricity is exported to the grid. However, an electricity network traditionally is not designed to deal with the 2- way energy transfer. Therefore, some technical issues may occur. For example in Queensland Australia, more than 30% of households used rooftop PV by the end of 2017. The famous Californian 2020 duck curve appeared often for a lot of communities from 2015 onwards. An over-voltage issue may result as the electricity flows from PV households back to the network. There are solutions to manage the over voltage issue, such as regulating PV inverter power factor, new voltage and energy control equipment at the electricity distributor level, re-conducting the electricity wires, demand side management, etc. There are often limitations and costs related to these solutions.

When electric networks are down, such as during the October 2019 California power shutoff, solar panels are often insufficient to fully provide power to a house or other structure, because they are designed to supply power to the grid, not directly to homes.

### Implication onto Electricity Bill Management and Energy Investment

There is no silver bullet in electricity or energy demand and bill management, because customers (sites) have different specific situations, e.g. different comfort/convenience needs, different electricity tariffs, or different usage patterns. Electricity tariff may have a few elements, such as daily access and metering charge, energy charge (based on kWh, MWh) or peak demand charge (e.g. a price for the highest 30min energy consumption in a month). PV is a promising option for reducing energy charge when electricity price is reasonably high and continuously increasing, such as in Australia and Germany. However for sites with peak demand charge in place, PV may be less attractive if peak demands mostly occur in the late afternoon to early evening, for example residential communities. Overall, energy investment is largely an economical decision and it is better to make investment decisions based on systematical evaluation of options in operational improvement, energy efficiency, onsite generation and energy storage.

## Floating Solar

Floating solar refers to a solar power production installation mounted on a structure that floats on a body of water, typically an artificial basin or a lake.

Two systems can be distinguished:

FPV or Floating photovoltaic: Uses photovoltaic panels mounted on the platform. Floating CSP or Floating concentrated solar power: uses mirrors that redirect the solar power to a tower.

## FPV

This technology has had a rapid growth on the renewable energy market since 2016 and in 2017 has overcome the 200 MW of installed power. The first 20 plants, of a few dozen of kWp have been built between 2008 and 2014 as reported in the MIRARCO paper that analyzed the birth of this technology.

Floating photovoltaic.

The installed power reached 1.1 GW in 2018. The costs for a floating system are 20-25% higher than for ground-mounted systems.

## Technology Features

There are several reasons for this development:

- No land occupancy: The main advantage of floating PV plants is that they do not take up any land, except the limited surfaces necessary for electric cabinet and grid connections. Their price is comparable with land based plants, but they provide a good way to avoid land consumption.
- Installation and decommissioning: Floating PV plants are more compact than land-based plants, their management is simpler and their construction and decommissioning straightforward. The main point is that no fixed structures exist like the foundations used for a land-based plant so their installation can be totally reversible.
- Water saving and water quality: The partial coverage of basins can reduce the water evaporation. This result depends on climate conditions and on the percentage of the covered surface. In arid climates such as Australia this is an important advantage since about 80% of the evaporation of the covered surface is saved and this means more than 20,000 $m^3$/year/ha. This is a very useful feature if the basin is used for irrigation purposes.
- Cooling: The floating structure allows the implementation of a simple cooling system. Cooling mechanism is natural but can also be active by generating a

water layer on the PV modules or using a submerged PV modules, the so-called SP2 (Submerged Photovoltaic Solar Panel). In these cases the global PV modules efficiency rises thanks to the absence of thermal drift, with a gain in energy harvesting up to 8-10%.

- Tracking: A large floating platform can be easily turned and can perform a vertical axis tracking: this can be done without wasting energy and without the need for a complex mechanical apparatus as in land-based PV plants. A floating PV plant equipped with a tracking system has a limited additional cost while the energy gain can range from 15 to 25%.
- Storage opportunity: The presence of water naturally suggests using gravity energy storage mainly in the coupling with hydroelectric basins. However other possibilities has been explored and in particular CAES systems have been suggested.
- Environment control: A parallel advantage is the containment of the algae bloom, a serious problem in industrialized countries. The partial coverage of the basins and the reduction of light on biological fouling just below the surface, together with active systems can solve this problem. This is only a part of the more general problem of managing a water basin generated by industrial activities or polluted by them. See for example the mining managing.
- Efficiency improvement: Many studies claim that there is a significant improvement in efficiency putting solar panels over water. These studies are not conclusive and differ in their conclusion. The energy gain reported range from 5 to 15%.

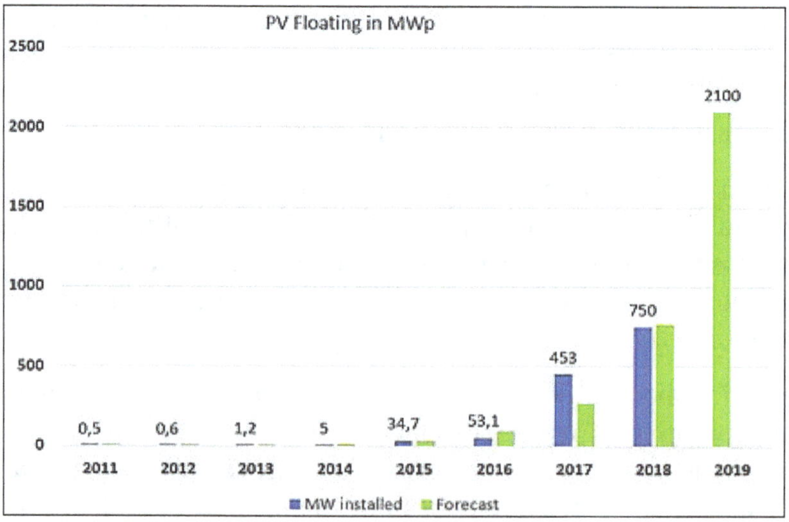


Installed capacity worldwide in MW.

American, Danish, French, Italian and Japanese nationals were the first to register patents for floating solar. In Italy the first registered patent, regarding PV modules on water, goes back to February 2008.

The MIRARCO (Mining Innovation Rehabilitation and Applied Research Corporation Ontario, CANADA) research group quotes several solutions that were put forward in the years 2008-2011 and 2012-2014. Without being exhaustive the installations can be classified into three categories:

- PV plants constituted by modules mounted on pontoons.
- PV modules mounted on rafts built in plastic and galvanized steel.
- PV modules mounted on rafts, fully in plastic.

It is Impossible to give a detailed analysis of the many small PV floating plants built in the first 10 years. The plot here below is based on data taken from the web for FPV with more than 500 kW of power. In the Asian Clean Energy Summit in Singapore two numbers where quoted by the World- Bank Group: 453 MWp for installation in 2017 and a forecast of 750 MWp for 2018.

The following graph shows the growth of solar floating installations globally from the beginning.

### Floating CSP

Floating CSP has similar advantages to floating photovoltaics.

## Photovoltaics

Method of generating electrical power by converting solar radiation into direct current electricity.

Photovoltaics (PV) is the conversion of light into electricity using semiconducting materials that exhibit the photovoltaic effect, a phenomenon studied in physics, photochemistry, and electrochemistry.

PV has become the cheapest source of electrical power in regions with a high potential, with price bids as low as 0.01567 US$/kWh in 2020. Panel prices have dropped by the factor of 10 within a decade. This competitiveness opens the path to a global transition to sustainable energy which would be required help to mitigate global warming. The situation is urgent: The emissions budget for CO2 to meet the 1.5 degree target would be used up in 2028 if emissions remain on the current level. However, the use of PV as a main source requires energy storage systems or global distribution by High-voltage direct current power lines causing additional costs.

A photovoltaic system employs solar modules, each comprising a number of solar cells, which generate electrical power. PV installations may be ground-mounted, rooftop mounted, wall mounted or floating. The mount may be fixed or use a solar tracker to follow the sun across the sky.

The Solar Settlement, a sustainable housing community project.

Photovoltaic SUDI shade is an autonomous and mobile station that provides energy for electric cars using solar energy.

Solar PV has specific advantages as an energy source: once installed, its operation generates no pollution and no greenhouse gas emissions, it shows simple scalability in respect of power needs and silicon has large availability in the Earth's crust.

Photovoltaic systems have long been used in specialized applications as stand-alone installations and grid-connected PV systems have been in use since the 1990s. Photovoltaic modules were first mass-produced in 2000, when German environmentalists and the Eurosolar organization got government funding for a ten thousand roof program.

Solar panels on the International Space Station.

Advances in technology and increased manufacturing scale have in any case reduced the cost, increased the reliability, and increased the efficiency of photovoltaic installations. Net metering and financial incentives, such as preferential feed-in tariffs for solar-generated electricity, have supported solar PV installations in many countries. More than 100 countries now use solar PV.

After hydro and wind powers, PV is the third renewable energy source in terms of global capacity. The International Energy Agency expects a growth by 700 - 880 GW from

2019 to 2024. According to the forecast, PV could become the technology with the largest installed capacity by the mid 2020s.

In 2018, worldwide installed PV capacity increased to more than 515 gigawatts (GW) covering approximately two percent of global electricity demand.

With current technology, photovoltaics recoups the energy needed to manufacture them in 1.5 years in Southern Europe and 2.5 years in Northern Europe.

## Solar Cells

Solar cells generate electricity directly from sunlight.

Photovoltaics are best known as a method for generating electric power by using solar cells to convert energy from the sun into a flow of electrons by the photovoltaic effect.

Solar cells produce direct current electricity from sunlight which can be used to power equipment or to recharge a battery. The first practical application of photovoltaics was to power orbiting satellites and other spacecraft, but today the majority of photovoltaic modules are used for grid-connected systems for power generation. In this case an inverter is required to convert the DC to AC. There is still a smaller market for stand alone systems for remote dwellings, boats, recreational vehicles, electric cars, roadside emergency telephones, remote sensing, and cathodic protection of pipelines.

Photovoltaic power generation employs solar modules composed of a number of solar cells containing a semiconductor material. Copper solar cables connect modules (module cable), arrays (array cable), and sub-fields. Because of the growing demand for renewable energy sources, the manufacturing of solar cells and photovoltaic arrays has advanced considerably in recent years.

Solar photovoltaic power generation has long been seen as a clean energy technology which draws upon the planet's most plentiful and widely distributed renewable energy source – the sun. Cells require protection from the environment and are usually packaged tightly in solar modules.

Photovoltaic module power is measured under standard test conditions (STC) in "$W_p$" (watts peak). The actual power output at a particular place may be less than or greater than this rated value, depending on geographical location, time of day, weather conditions, and other factors. Solar photovoltaic array capacity factors are typically under 25%, which is lower than many other industrial sources of electricity.

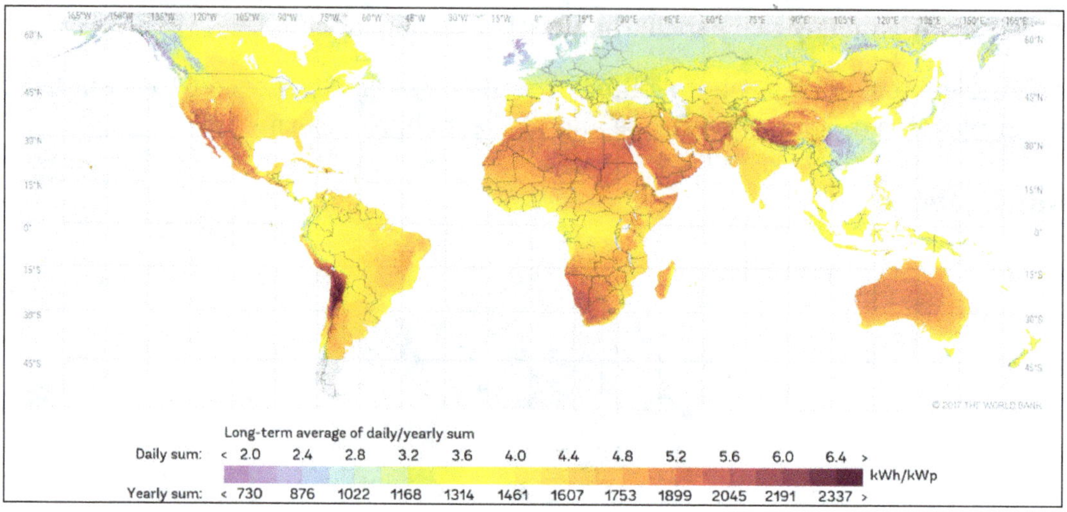


Photovoltaic power potential map estimates, how many kWh of electricity can be produced from a 1 kWp free-standing c-Si modules, optimally inclined towards the Equator. The resulting long-term average (daily or yearly) is calculated based on the time-series weather data of at least recent 10 years.

## Current Developments

For best performance, terrestrial PV systems aim to maximize the time they face the sun. Solar trackers achieve this by moving PV modules to follow the sun. The increase can be by as much as 20% in winter and by as much as 50% in summer. Static mounted systems can be optimized by analysis of the sun path. PV modules are often set to latitude tilt, an angle equal to the latitude, but performance can be improved by adjusting the angle for summer or winter. Generally, as with other semiconductor devices, temperatures above room temperature reduce the performance of photovoltaic modules.

A number of solar modules may also be mounted vertically above each other in a tower, if the zenith distance of the Sun is greater than zero, and the tower can be turned horizontally as a whole and each module additionally around a horizontal axis. In such a tower the modules can follow the Sun exactly. Such a device may be described as a ladder mounted on a turnable disk. Each step of that ladder is the middle axis of a rectangular solar panel. In case the zenith distance of the Sun reaches zero, the "ladder" may be rotated to the north or the south to avoid a solar module producing a shadow on a lower one. Instead of an exactly vertical tower one can choose a tower with an axis directed to the polar star, meaning that it is parallel to the rotation axis of the Earth. In this case the angle between the axis and the Sun is always larger than 66 degrees.

During a day it is only necessary to turn the panels around this axis to follow the Sun. Installations may be ground-mounted (and sometimes integrated with farming and grazing) or built into the roof or walls of a building (building-integrated photovoltaics).

Another recent development involves the material content of solar cells. University labs at Oxford, Stanford, and elsewhere are experimenting with replacing crystalline silicon in photovoltaic solar cells with perovskite, a mineral with semiconductor properties. Researchers at these labs have quickly achieved cell efficiencies of 20-25%, matching or exceeding those of silicon solar cells.

## Efficiency

The electrical efficiency of a PV cell is a physical property which represents how much electrical power a cell can produce for a given Solar irradiance. The basic expression for maximum efficiency of a photovoltaic cell is given by the ratio of output power to the incident solar power (radiation flux times area),

$$\eta = \frac{P_{max}}{E \cdot A_{cell}}.$$

The efficiency is measured under ideal laboratory conditions and represents the maximum achievable efficiency of the PV cell or module. Actual efficiency is influenced by temperature, irradiance and spectrum.

The most efficient type of solar cell to date is a multi-junction concentrator solar cell with an efficiency of 46.0% produced by Fraunhofer ISE in December 2014. The highest efficiencies achieved without concentration include a material by Sharp Corporation at 35.8% using a proprietary triple-junction manufacturing technology in 2009, and Boeing Spectrolab (40.7% also using a triple-layer design). The US-based specialty gallium arsenide (GaAs) PV manufacturer Alta Devices produces commercial cells with 26% efficiency claiming to have "the world's most efficient solar" single-junction cell dedicated to flexible and lightweight applications. For Silicon solar cell, the US company SunPower remains the leader with a certified module efficiency of 22.8%, well above the market average of 15–18%. However, competitor companies are catching up like the South Korean conglomerate LG (21.7% efficiency) or the Norwegian REC Group (21.7% efficiency).

There is an ongoing effort to increase the conversion efficiency of PV cells and modules, primarily for competitive advantage. In order to increase the efficiency of solar cells, it is important to choose a semiconductor material with an appropriate band gap that matches the solar spectrum. This will enhance the electrical and optical properties. Improving the method of charge collection is also useful for increasing the efficiency. There are several groups of materials that are being developed. Ultrahigh-efficiency devices (η>30%) are made by using GaAs and GaInP2 semiconductors with multijunction tandem cells. High-quality, single-crystal silicon materials are used to achieve high-efficiency, low cost cells (η>20%).

Recent developments in Organic photovoltaic cells (OPVs) have made significant advancements in power conversion efficiency from 3% to over 15% since their introduction in the 1980s. To date, the highest reported power conversion efficiency ranges from 6.7% to 8.94% for small molecule, 8.4%–10.6% for polymer OPVs, and 7% to 21% for perovskite OPVs. OPVs are expected to play a major role in the PV market. Recent improvements have increased the efficiency and lowered cost, while remaining environmentally-benign and renewable.

Perovskite cells are also rapidly (as of 2019) progressing, with top efficiencies increasing from 13% to 23,7 % for non stabilized cells and up to 28% for perovskite/Si tandem monolithic cells between 2012 and 2018.

Several companies have begun embedding power optimizers into PV modules called smart modules. These modules perform maximum power point tracking (MPPT) for each module individually, measure performance data for monitoring, and provide additional safety features. Such modules can also compensate for shading effects, wherein a shadow falling across a section of a module causes the electrical output of one or more strings of cells in the module to decrease.

One of the major causes for the decreased performance of cells is overheating. The efficiency of a solar cell declines by about 0.5% for every 1 degree Celsius increase in temperature. This means that a 100 degree increase in surface temperature could decrease the efficiency of a solar cell by about half. Self-cooling solar cells are one solution to this problem. Rather than using energy to cool the surface, pyramid and cone shapes can be formed from silica, and attached to the surface of a solar panel. Doing so allows visible light to reach the solar cells, but reflects infrared rays (which carry heat).

## Growth

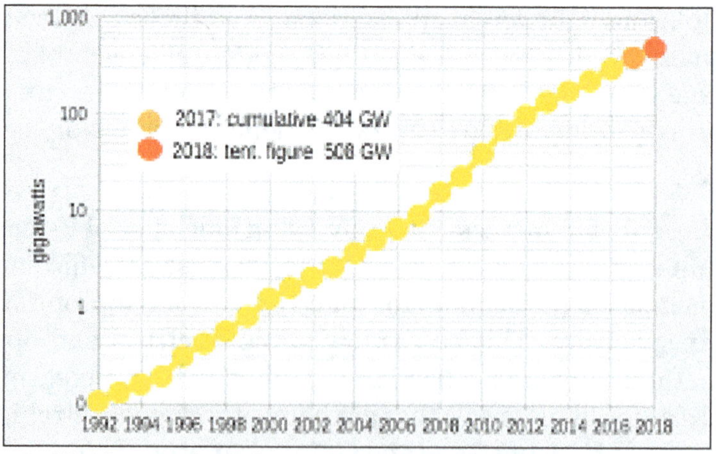


Worldwide growth of photovoltaics on a semi-log plot since 1992.

Solar photovoltaics is growing rapidly and worldwide installed capacity reached about 515 gigawatts (GW) by 2018. The total power output of the world's PV capacity in a calendar year now beyond 500 TWh of electricity. This represents 2% of worldwide electricity demand. More than 100 countries use solar PV. China is followed by the United States and Japan, while installations in Germany, once remains the world's largest producer, have been slowing down.

In 2017 a study in Science estimated that by 2030 global PV installed capacities will be between 3,000 and 10,000 GW. The EPIA/Greenpeace Solar Generation Paradigm Shift Scenario (formerly called Advanced Scenario) from 2010 shows that by the year 2030, 1,845 GW of PV systems could be generating approximately 2,646 TWh/year of electricity around the world. Combined with energy use efficiency improvements, this would represent the electricity needs of more than 9% of the world's population. By 2050, over 20% of all electricity could be provided by photovoltaics.

Michael Liebreich, from Bloomberg New Energy Finance, anticipates a tipping point for solar energy. The costs of power from wind and solar are already below those of conventional electricity generation in some parts of the world, as they have fallen sharply and will continue to do so. He also asserts, that the electrical grid has been greatly expanded worldwide, and is ready to receive and distribute electricity from renewable sources. In addition, worldwide electricity prices came under strong pressure from renewable energy sources, that are, in part, enthusiastically embraced by consumers.

Deutsche Bank sees a "second gold rush" for the photovoltaic industry to come. Grid parity has already been reached in at least 19 markets by January 2014. Photovoltaics will prevail beyond feed-in tariffs, becoming more competitive as deployment increases and prices continue to fall.

In June 2014 Barclays downgraded bonds of U.S. utility companies. Barclays expects more competition by a growing self-consumption due to a combination of decentralized PV-systems and residential electricity storage. This could fundamentally change the utility's business model and transform the system over the next ten years, as prices for these systems are predicted to fall.

## Environmental Impacts of Photovoltaic Technologies

## Types of Impacts

While solar photovoltaic (PV) cells are promising for clean energy production, their deployment is hindered by production costs, material availability, and toxicity. Data required to investigate their impact are sometimes affected by a rather large amount of uncertainty. The values of human labor and water consumption, for example, are not precisely assessed due to the lack of systematic and accurate analyses in the scientific literature.

Life cycle assessment (LCA) is one method of determining environmental impacts from PV. Many studies have been done on the various types of PV including first generation, second generation, and third generation. Usually these PV LCA studies select a cradle to gate system boundary because often at the time the studies are conducted, it is a new technology not commercially available yet and their required balance of system components and disposal methods are unknown.

A traditional LCA can look at many different impact categories ranging from global warming potential, ecotoxicity, human toxicity, water depletion, and many others.

Most LCAs of PV have focused on two categories: carbon dioxide equivalents per kWh and energy pay-back time (EPBT). The EPBT is defined as " the time needed to compensate for the total renewable- and non-renewable- primary energy required during the life cycle of a PV system". A 2015 review of EPBT from first and second generation PV suggested that there was greater variation in embedded energy than in efficiency of the cells implying that it was mainly the embedded energy that needs to reduce to have a greater reduction in EPBT. One difficulty in determining impacts due to PV is to determine if the wastes are released to the air, water, or soil during the manufacturing phase. Research is underway to try to understand emissions and releases during the lifetime of PV systems.

## Impacts from First Generation PV

Crystalline silicon modules are the most extensively studied PV type in terms of LCA since they are the most commonly used. Mono-crystalline silicon photovoltaic systems (mono-si) have an average efficiency of 14.0%. The cells tend to follow a structure of front electrode, anti-reflection film, n-layer, p-layer, and back electrode, with the sun hitting the front electrode. EPBT ranges from 1.7 to 2.7 years. The cradle to gate of $CO_2$-eq/kWh ranges from 37.3 to 72.2 grams.

Techniques to produce multi-crystalline silicon (multi-si) photovoltaic cells are simpler and cheaper than mono-si, however tend to make less efficient cells, an average of 13.2%. EPBT ranges from 1.5 to 2.6 years. The cradle to gate of $CO_2$-eq/kWh ranges from 28.5 to 69 grams. Some studies have looked beyond EPBT and GWP to other environmental impacts. In one such study, conventional energy mix in Greece was compared to multi-si PV and found a 95% overall reduction in impacts including carcinogens, eco-toxicity, acidification, eutrophication, and eleven others.

## Impacts from Second Generation

Cadmium telluride (CdTe) is one of the fastest-growing thin film based solar cells which are collectively known as second generation devices. This new thin film device also shares similar performance restrictions (Shockley-Queisser efficiency limit) as conventional Si devices but promises to lower the cost of each device by both reducing material and energy consumption during manufacturing. Today the global market share of CdTe is 5.4%, up from 4.7% in 2008. This technology's highest power conversion efficiency is 21%. The cell structure includes glass substrate (around 2 mm), transparent conductor layer, CdS buffer layer (50–150 nm), CdTe absorber and a metal contact layer.

CdTe PV systems require less energy input in their production than other commercial PV systems per unit electricity production. The average $CO_2$-eq/kWh is around 18 grams (cradle to gate). CdTe has the fastest EPBT of all commercial PV technologies, which varies between 0.3 and 1.2 years.

Copper Indium Gallium Diselenide (CIGS) is a thin film solar cell based on the copper indium diselenide (CIS) family of chalcopyrite semiconductors. CIS and CIGS are often used interchangeably within the CIS/CIGS community. The cell structure includes soda lime glass as the substrate, Mo layer as the back contact, CIS/CIGS as the absorber layer, cadmium sulfide (CdS) or Zn (S,OH)x as the buffer layer, and ZnO:Al as the front contact. CIGS is approximately 1/100th the thickness of conventional silicon solar cell technologies. Materials necessary for assembly are readily available, and are less costly per watt of solar cell. CIGS based solar devices resist performance degradation over time and are highly stable in the field.

Reported global warming potential impacts of CIGS range from 20.5 – 58.8 grams $CO_2$-eq/kWh of electricity generated for different solar irradiation (1,700 to 2,200 kWh/$m^2$/y) and power conversion efficiency (7.8 – 9.12%). EPBT ranges from 0.2 to 1.4 years, while harmonized value of EPBT was found 1.393 years. Toxicity is an issue within the buffer layer of CIGS modules because it contains cadmium and gallium. CIS modules do not contain any heavy metals.

## Impacts from Third Generation

Third-generation PVs are designed to combine the advantages of both the first and second generation devices and they do not have Shockley-Queisser limit, a theoretical limit for first and second generation PV cells. The thickness of a third generation device is less than 1 μm.

One emerging alternative and promising technology is based on an organic-inorganic hybrid solar cell made of methylammonium lead halide perovskites. Perovskite PV cells have progressed rapidly over the past few years and have become one of the most attractive areas for PV research. The cell structure includes a metal back contact (which can be made of Al, Au or Ag), a hole transfer layer (spiro-MeOTAD, P3HT, PTAA, CuSCN, CuI, or NiO), and absorber layer ($CH_3NH_3PbIxBr_3$-x, $CH_3NH_3PbIxCl_3$-x or $CH_3NH_3PbI_3$), an electron transport layer (TiO, ZnO, $Al_2O_3$ or $SnO_2$) and a top contact layer (fluorine doped tin oxide or tin doped indium oxide).

There are a limited number of published studies to address the environmental impacts of perovskite solar cells. The major environmental concern is the lead used in the absorber layer. Due to the instability of perovskite cells lead may eventually be exposed to fresh water during the use phase. These LCA studies looked at human and ecotoxicity of perovskite solar cells and found they were surprisingly low and may not be an environmental issue. Global warming potential of perovskite PVs were found to be in the range of 24–1500 grams $CO_2$-eq/kWh electricity production. Similarly, reported EPBT of the published paper range from 0.2 to 15 years. The large range of reported values highlight the uncertainties associated with these studies. Celik et al. critically discussed the assumptions made in perovskite PV LCA studies.

Two new promising thin film technologies are copper zinc tin sulfide ($Cu_2ZnSnS_4$ or CZTS), zinc phosphide ($Zn_3P_2$) and single-walled carbon nano-tubes (SWCNT). These thin films are currently only produced in the lab but may be commercialized in the future. The manufacturing of CZTS and ($Zn_3P_2$) processes are expected to be similar to those of current thin film technologies of CIGS and CdTe, respectively. While the absorber layer of SWCNT PV is expected to be synthesized with CoMoCAT method. by Contrary to established thin films such as CIGS and CdTe, CZTS, $Zn_3P_2$, and SWCNT PVs are made from earth abundant, nontoxic materials and have the potential to produce more electricity annually than the current worldwide consumption. While CZTS and $Zn_3P_2$ offer good promise for these reasons, the specific environmental implications of their commercial production are not yet known. Global warming potential of CZTS and $Zn_3P_2$ were found 38 and 30 grams $CO_2$-eq/kWh while their corresponding EPBT were found 1.85 and 0.78 years, respectively. Overall, CdTe and $Zn_3P_2$ have similar environmental impacts but can slightly outperform CIGS and CZTS. Celik et al. performed the first LCA study on environmental impacts of SWCNT PVs, including a laboratory-made 1% efficient device and an aspirational 28% efficient four-cell tandem device and interpreted the results by using mono-Si as a reference point. the results show that compared to monocrystalline Si (mono-Si), the environmental impacts from 1% SWCNT was ~18 times higher due mainly to the short lifetime of three years. However, even with the same short lifetime, the 28% cell had lower environmental impacts than mono-Si.

Organic and polymer photovoltaic (OPV) are a relatively new area of research. The tradition OPV cell structure layers consist of a semi-transparent electrode, electron blocking layer, tunnel junction, holes blocking layer, electrode, with the sun hitting the transparent electrode. OPV replaces silver with carbon as an electrode material lowering manufacturing cost and making them more environmentally friendly. OPV are flexible, low weight, and work well with roll-to roll manufacturing for mass production. OPV uses “only abundant elements coupled to an extremely low embodied energy through very low processing temperatures using only ambient processing conditions on simple printing equipment enabling energy pay-back times”. Current efficiencies range from 1–6.5%, however theoretical analyses show promise beyond 10% efficiency.

Many different configurations of OPV exist using different materials for each layer. OPV technology rivals existing PV technologies in terms of EPBT even if they currently present a shorter operational lifetime. A 2013 study analyzed 12 different configurations all with 2% efficiency, the EPBT ranged from 0.29–0.52 years for 1 $m^2$ of PV. The average $CO_2$-eq/kWh for OPV is 54.922 grams.

## Economics

There have been major changes in the underlying costs, industry structure and market prices of solar photovoltaics technology, over the years, and gaining a coherent picture of the shifts occurring across the industry value chain globally is a challenge. This is due to: "the rapidity of cost and price changes, the complexity of the PV supply chain, which involves a large number of manufacturing processes, the balance of system (BOS) and installation costs associated with complete PV systems, the choice of different distribution channels, and differences between regional markets within which PV is being deployed". Further complexities result from the many different policy support initiatives that have been put in place to facilitate photovoltaics commercialisation in various countries.

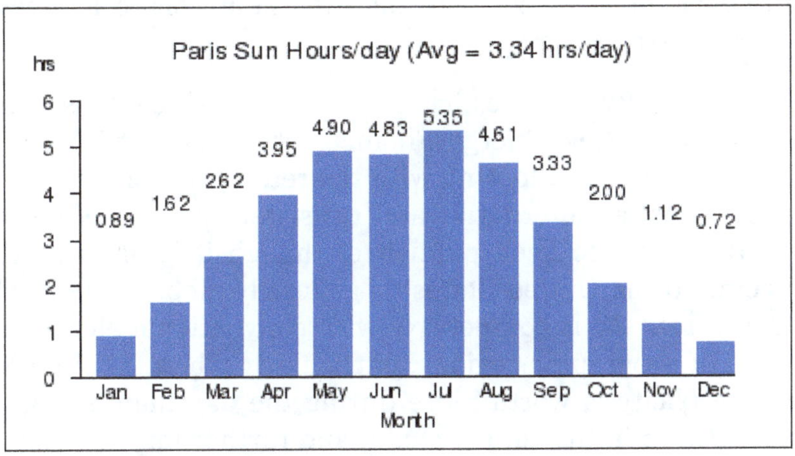


The PV industry has seen dramatic drops in module prices since 2008. In late 2011, factory-gate prices for crystalline-silicon photovoltaic modules dropped below the $1.00/W mark. The $1.00/W installed cost, is often regarded in the PV industry as marking the achievement of grid parity for PV. Technological advancements, manufacturing process improvements, and industry re-structuring, mean that further price reductions are likely in coming years. As of 2017 power-purchase agreement prices for solar farms below $0.05/kWh are common in the United States and the lowest bids in several international countries were about $0.03/kWh.

Financial incentives for photovoltaics, such as feed-in tariffs, have often been offered to electricity consumers to install and operate solar-electric generating systems. Government has sometimes also offered incentives in order to encourage the PV industry to achieve the economies of scale needed to compete where the cost of PV-generated electricity is above the cost from the existing grid. Such policies are implemented to promote national or territorial energy independence, high tech job creation and reduction of carbon dioxide emissions which cause climate change. Due to economies of scale solar panels get less costly as people use and buy more—as manufacturers increase production to meet demand, the cost and price is expected to drop in the years to come.

Solar cell efficiencies vary from 6% for amorphous silicon-based solar cells to 44.0% with multiple-junction concentrated photovoltaics. Solar cell energy conversion efficiencies for commercially available photovoltaics are around 14–22%. Concentrated photovoltaics (CPV) may reduce cost by concentrating up to 1,000 suns (through magnifying lens) onto a smaller sized photovoltaic cell. However, such concentrated solar power requires sophisticated heat sink designs, otherwise the photovoltaic cell overheats, which reduces its efficiency and life. To further exacerbate the concentrated cooling design, the heat sink must be passive, otherwise the power required for active cooling would reduce the overall efficiency and economy.

Crystalline silicon solar cell prices have fallen from $76.67/Watt in 1977 to an estimated $0.74/Watt in 2013. This is seen as evidence supporting Swanson's law, an observation similar to the famous Moore's Law that states that solar cell prices fall 20% for every doubling of industry capacity.

As of 2011, the price of PV modules has fallen by 60% since the summer of 2008, according to Bloomberg New Energy Finance estimates, putting solar power for the first time on a competitive footing with the retail price of electricity in a number of sunny countries; an alternative and consistent price decline figure of 75% from 2007 to 2012 has also been published, though it is unclear whether these figures are specific to the United States or generally global. The levelised cost of electricity (LCOE) from PV is competitive with conventional electricity sources in an expanding list of geographic regions, particularly when the time of generation is included, as electricity is worth more during the day than at night. There has been fierce competition in the supply chain, and further improvements in the levelised cost of energy for solar lie ahead, posing a growing threat to the dominance of fossil fuel generation sources in the next few years. As time progresses, renewable energy technologies generally get cheaper, while fossil fuels generally get more expensive:

> The less solar power costs, the more favorably it compares to conventional power, and the more attractive it becomes to utilities and energy users around the globe. Utility-scale solar power can now be delivered in California at prices well below $100/MWh ($0.10/kWh) less than most other peak generators, even those running on low-cost natural gas. Lower solar module costs also stimulate demand from consumer markets where the cost of solar compares very favorably to retail electric rates.

As of 2011, the cost of PV has fallen well below that of nuclear power and is set to fall further. The average retail price of solar cells as monitored by the Solarbuzz group fell from $3.50/watt to $2.43/watt over the course of 2011.

> For large-scale installations, prices below $1.00/watt were achieved. A module price of 0.60 Euro/watt ($0.78/watt) was published for a large scale 5-year deal in April 2012.

By the end of 2012, the "best in class" module price had dropped to $0.50/watt, and was expected to drop to $0.36/watt by 2017.

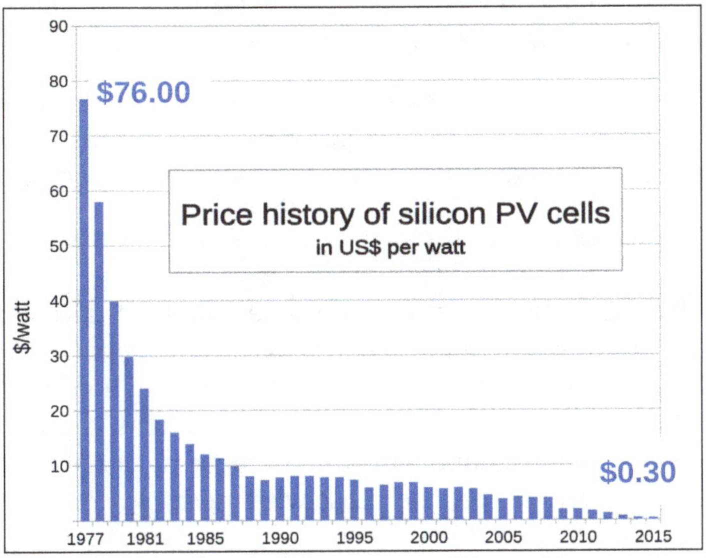


Price per watt history for conventional (c-Si) solar cells since 1977.

In many locations, PV has reached grid parity, which is usually defined as PV production costs at or below retail electricity prices (though often still above the power station prices for coal or gas-fired generation without their distribution and other costs). However, in many countries there is still a need for more access to capital to develop PV projects. To solve this problem securitization has been proposed and used to accelerate development of solar photovoltaic projects. For example, SolarCity offered, the first U.S. asset-backed security in the solar industry in 2013.

Photovoltaic power is also generated during a time of day that is close to peak demand (precedes it) in electricity systems with high use of air conditioning. Since large-scale PV operation requires back-up in the form of spinning reserves, its marginal cost of generation in the middle of the day is typically lowest, but not zero, when PV is generating electricity. This can be seen in figure of this paper: More generally, it is now evident that, given a carbon price of $50/ton, which would raise the price of coal-fired power by 5c/kWh, solar PV will be cost-competitive in most locations. The declining price of PV has been reflected in rapidly growing installations, totaling about 23 GW in 2011. Although some consolidation is likely in 2012, due to support cuts in the large markets of Germany and Italy, strong growth seems likely to continue for the rest of the decade. Already, by one estimate, total investment in renewables for 2011 exceeded investment in carbon-based electricity generation.

In the case of self consumption payback time is calculated based on how much electricity is not brought from the grid. Additionally, using PV solar power to charge DC batteries, as used in Plug-in Hybrid Electric Vehicles and Electric Vehicles, leads to greater efficiencies. Traditionally, DC generated electricity from solar PV must be converted to

AC for buildings, at an average 10% loss during the conversion. An additional efficiency loss occurs in the transition back to DC for battery driven devices and vehicles, and using various interest rates and energy price changes were calculated to find present values that range from $2,057 to $8,213 (analysis from 2009).

For example, in Germany with electricity prices of 0.25 euro/kWh and Insolation of 900 kWh/kW one $kW_p$ will save 225 euro per year and with installation cost of 1700 euro/$kW_p$ means that the system could pay back in less than 7 years (example for 2013).

## Manufacturing

Overall the manufacturing process of creating solar photovoltaics is simple in that it does not require the culmination of many complex or moving parts. Because of the solid state nature of PV systems they often have relatively long lifetimes, anywhere from 10 to 30 years. To increase electrical output of a PV system, the manufacturer must simply add more photovoltaic components and because of this economies of scale are important for manufacturers as costs decrease with increasing output.

While there are many types of PV systems known to be effective, crystalline silicon PV accounted for around 90% of the worldwide production of PV in 2013. Manufacturing silicon PV systems has several steps. First, polysilicon is processed from mined quartz until it is very pure (semi-conductor grade). This is melted down when small amounts of boron, a group III element, are added to make a p-type semiconductor rich in electron holes. Typically using a seed crystal, an ingot of this solution is grown from the liquid polycrystalline. The ingot may also be cast in a mold. Wafers of this semiconductor material are cut from the bulk material with wire saws, and then go through surface etching before being cleaned. Next, the wafers are placed into a phosphorus vapor deposition furnace which lays a very thin layer of phosphorus, a group V element, which creates an n-type semiconducting surface. To reduce energy losses, an anti-reflective coating is added to the surface, along with electrical contacts. After finishing the cell, cells are connected via electrical circuit according to the specific application and prepared for shipping and installation.

Crystalline silicon photovoltaics are only one type of PV, and while they represent the majority of solar cells produced currently there are many new and promising technologies that have the potential to be scaled up to meet future energy needs. As of 2018, crystalline silicon cell technology serves as the basis for several PV module types, including monocrystalline, multicrystalline, mono PERC, and bifacial.

Another newer technology, Thin film PV, are manufactured by depositing semiconducting layers on substrate in vacuum. The substrate is often glass or stainless-steel, and these semiconducting layers are made of many types of materials including cadmium telluride (CdTe), copper indium diselenide (CIS), copper indium gallium diselenide (CIGS), and amorphous silicon (a-Si). After being deposited onto the substrate the semiconducting layers are separated and connected by electrical circuit by laser-scribing. Thin film photovoltaics now make up around 20% of the overall production of

PV because of the reduced materials requirements and cost to manufacture modules consisting of Thin films as compared to silicon-based wafers.

Other emerging PV technologies include organic, dye-sensitized, quantum-dot, and Perovskite photovoltaics. OPVs fall into the Thin film category of manufacturing, and typically operate around the 12% efficiency range which is lower than the 12–21% typically seen by silicon based PVs. Because organic photovoltaics require very high purity and are relatively reactive they must be encapsulated which vastly increases cost of manufacturing and meaning that they are not feasible for large scale up. Dye-sensitized PVs are similar in efficiency to OPVs but are significantly easier to manufacture. However these dye-sensitized photovoltaics present storage problems because the liquid electrolyte is toxic and can potentially permeate the plastics used in the cell. Quantum dot solar cells are quantum dot sensitized DSSCs and are solution processed meaning they are potentially scalable, but currently they peak at 12% efficiency. Perovskite solar cells are a very efficient solar energy converter and have excellent optoelectric properties for photovoltaic purposes, but they are expensive and difficult to manufacture.

## Applications

### Photovoltaic Systems

A photovoltaic system, or solar PV system is a power system designed to supply usable solar power by means of photovoltaics. It consists of an arrangement of several components, including solar panels to absorb and directly convert sunlight into electricity, a solar inverter to change the electric current from DC to AC, as well as mounting, cabling and other electrical accessories. PV systems range from small, roof-top mounted or building-integrated systems with capacities from a few to several tens of kilowatts, to large utility-scale power stations of hundreds of megawatts. Nowadays, most PV systems are grid-connected, while stand-alone systems only account for a small portion of the market.

Rooftop and building integrated systems: Photovoltaic arrays are often associated with buildings: either integrated into them, mounted on them or mounted nearby on the ground. Rooftop PV systems are most often retrofitted into existing buildings, usually mounted on top of the existing roof structure or on the existing walls. Alternatively, an array can be located separately from the building but connected by cable to supply power for the building. Building-integrated photovoltaics (BIPV) are increasingly incorporated into the roof or walls of new domestic and industrial buildings as a principal or ancillary source of electrical power. Roof tiles with integrated PV cells are sometimes used as well. Provided there is an open gap in which air can circulate, rooftop mounted solar panels can provide a passive cooling effect on buildings during the day and also keep accumulated heat in at night. Typically, residential rooftop systems have small capacities of around 5–10 kW, while commercial rooftop systems often amount to several hundreds of kilowatts. Although rooftop systems are much smaller than ground-mounted utility-scale power plants, they account for most of the worldwide installed capacity.

Rooftop PV on half-timbered house.

Concentrator photovoltaics: Concentrator photovoltaics (CPV) is a photovoltaic technology that contrary to conventional flat-plate PV systems uses lenses and curved mirrors to focus sunlight onto small, but highly efficient, multi-junction (MJ) solar cells. In addition, CPV systems often use solar trackers and sometimes a cooling system to further increase their efficiency. Ongoing research and development is rapidly improving their competitiveness in the utility-scale segment and in areas of high solar insolation.

Photovoltaic thermal hybrid solar collector: Photovoltaic thermal hybrid solar collector (PVT) are systems that convert solar radiation into thermal and electrical energy. These systems combine a solar PV cell, which converts sunlight into electricity, with a solar thermal collector, which captures the remaining energy and removes waste heat from the PV module. The capture of both electricity and heat allow these devices to have higher exergy and thus be more overall energy efficient than solar PV or solar thermal alone.

Power stations: Many utility-scale solar farms have been constructed all over the world. As of 2015, the 579-megawatt ($MW_{AC}$) Solar Star is the world's largest photovoltaic power station, followed by the Desert Sunlight Solar Farm and the Topaz Solar Farm, both with a capacity of 550 $MW_{AC}$, constructed by US-company First Solar, using CdTe modules, a Thin film PV technology. All three power stations are located in the Californian desert. Many solar farms around the world are integrated with agriculture and some use innovative solar tracking systems that follow the sun's daily path across the sky to generate more electricity than conventional fixed-mounted systems. There are no fuel costs or emissions during operation of the power stations.

Rural electrification: Developing countries where many villages are often more than five kilometers away from grid power are increasingly using photovoltaics. In remote locations in India a rural lighting program has been providing solar powered LED lighting to replace kerosene lamps. The solar powered lamps were sold at about the cost of a few months' supply of kerosene. Cuba is working to provide solar power for areas that are off grid. More complex applications of off-grid solar energy use include 3D printers. RepRap 3D printers have been solar powered with photovoltaic technology, which enables distributed manufacturing for sustainable development. These are areas where the social costs and benefits offer an excellent case for going solar, though the lack of profitability has relegated such endeavors to humanitarian efforts. However, in 1995 solar rural electrification projects had been found to be difficult to sustain due to unfavorable economics, lack of technical support, and a legacy of ulterior motives of north-to-south technology transfer.

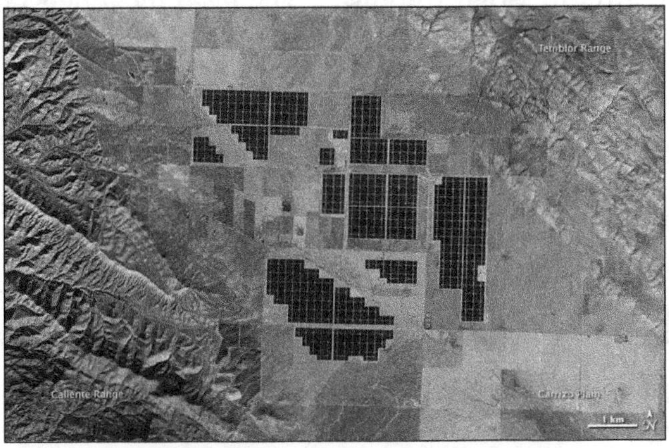

Satellite image of the Topaz Solar Farm.

Standalone systems: Until a decade or so ago, PV was used frequently to power calculators and novelty devices. Improvements in integrated circuits and low power liquid crystal displays make it possible to power such devices for several years between battery changes, making PV use less common. In contrast, solar powered remote fixed devices have seen increasing use recently in locations where significant connection cost makes grid power prohibitively expensive. Such applications include solar lamps, water pumps, parking meters, emergency telephones, trash compactors, temporary traffic signs, charging stations, and remote guard posts and signals.

Floating solar: Where land may be limited, PV can be deployed as floating solar. In May 2008, the Far Niente Winery in Oakville, CA pioneered the world's first "floatovoltaic" system by installing 994 photovoltaic solar panels onto 130 pontoons and floating them on the winery's irrigation pond. The floating system generates about 477 kW of peak output and when combined with an array of cells located adjacent to the pond is able to fully offset the winery's electricity consumption. The primary benefit of a floating system is that it avoids the need to sacrifice valuable land area that could be used for another purpose. In the case of the Far Niente Winery, the floating system saved three-quarters of an acre that would have been required for a land-based system. That land area can instead be used for agriculture. Another benefit of a floating solar system is that the panels are kept at a lower temperature than they would be on land, leading to a higher efficiency of solar energy conversion. The floating panels also reduce the amount of water lost through evaporation and inhibit the growth of algae.

In transport: PV has traditionally been used for electric power in space. PV is rarely used to provide motive power in transport applications, but is being used increasingly to provide auxiliary power in boats and cars. Some automobiles are fitted with solar-powered air conditioning to limit interior temperatures on hot days. A self-contained solar vehicle would have limited power and utility, but a solar-charged electric vehicle allows use of solar power for transportation. Solar-powered cars, boats and airplanes have been demonstrated, with the most practical and likely of these being solar cars. The Swiss solar aircraft, Solar Impulse 2, achieved the longest non-stop solo flight in history and completed the first solar-powered aerial circumnavigation of the globe in 2016.

Solar Impulse 2, a solar aircraft.

Telecommunication and signalling: Solar PV power is ideally suited for telecommunication applications such as local telephone exchange, radio and TV broadcasting, microwave and other forms of electronic communication links. This is because, in most telecommunication application, storage batteries are already in use and the electrical system is basically DC. In hilly and mountainous terrain, radio and TV signals may not reach as they get blocked or reflected back due to undulating terrain. At these locations, low power transmitters (LPT) are installed to receive and retransmit the signal for local population.

Spacecraft applications: Solar panels on spacecraft are usually the sole source of power to run the sensors, active heating and cooling, and communications. A battery stores this energy for use when the solar panels are in shadow. In some, the power is also used for spacecraft propulsion—electric propulsion. Spacecraft were one of the earliest applications of photovoltaics, starting with the silicon solar cells used on the Vanguard 1 satellite, launched by the US in 1958. Since then, solar power has been used on missions ranging from the MESSENGER probe to Mercury, to as far out in the solar system as the Juno probe to Jupiter. The largest solar power system flown in space is the electrical system of the International Space Station. To increase the power generated per kilogram, typical spacecraft solar panels use high-cost, high-efficiency, and close-packed rectangular multi-junction solar cells made of gallium arsenide (GaAs) and other semiconductor materials.

Specialty Power Systems: Photovoltaics may also be incorporated as energy conversion devices for objects at elevated temperatures and with preferable radiative emissivities such as heterogeneous combustors.

## Advantages

The 122 PW of sunlight reaching the Earth's surface is plentiful—almost 10,000 times more than the 13 TW equivalent of average power consumed in 2005 by humans. This abundance leads to the suggestion that it will not be long before solar energy will become the world's primary energy source. Additionally, solar electric generation has the highest power density (global mean of 170 W/m$^2$) among renewable energies.

Solar power is pollution-free during use, which enables it to cut down on pollution when it is substituted for other energy sources. For example, MIT estimated that 52,000 people per year die prematurely in the U.S. from coal-fired power plant pollution and all but one of these deaths could be prevented from using PV to replace coal. Production end-wastes and emissions are manageable using existing pollution controls. End-of-use recycling technologies are under development and policies are being produced that encourage recycling from producers.

PV installations could ideally operate for 100 years or even more with little maintenance or intervention after their initial set-up, so after the initial capital cost of building any solar power plant, operating costs are extremely low compared to existing power technologies.

Grid-connected solar electricity can be used locally thus reducing transmission/distribution losses.

Compared to fossil and nuclear energy sources, very little research money has been invested in the development of solar cells, so there is considerable room for improvement. Nevertheless, experimental high efficiency solar cells already have efficiencies of over 40% in case of concentrating photovoltaic cells and efficiencies are rapidly rising while mass-production costs are rapidly falling.

In some states of the United States, much of the investment in a home-mounted system may be lost if the homeowner moves and the buyer puts less value on the system than the seller. The city of Berkeley developed an innovative financing method to remove this limitation, by adding a tax assessment that is transferred with the home to pay for the solar panels. Now known as PACE, Property Assessed Clean Energy, 30 U.S. states have duplicated this solution.

There is evidence, at least in California, that the presence of a home-mounted solar system can actually increase the value of a home. According to a paper published in April 2011 by the Ernest Orlando Lawrence Berkeley National Laboratory titled An Analysis of the Effects of Residential Photovoltaic Energy Systems on Home Sales Prices in California:

> The research finds strong evidence that homes with PV systems in California have sold for a premium over comparable homes without PV systems. More specifically, estimates for average PV premiums range from approximately $3.9 to $6.4 per installed watt (DC) among a large number of different model specifications, with most models coalescing near $5.5/watt. That value corresponds to a premium of approximately $17,000 for a relatively new 3,100 watt PV system (the average size of PV systems in the study).

## Limitations

- Pollution and Energy in Production: PV has been a well-known method of generating clean, emission-free electricity. PV systems are often made of PV modules and inverter (changing DC to AC). PV modules are mainly made of PV cells, which has no fundamental difference from the material used for making computer chips. The process of producing PV cells is energy-intensive and involves highly poisonous and environmentally toxic chemicals. There are a few PV manufacturing plants around the world producing PV modules with energy produced from PV. This counteractive measure considerably reduces the carbon footprint of the manufacturing process of PV cells. Management of the chemicals used and produced during the manufacturing process is subject to the factories' local laws and regulations.
- Impact on Electricity Network: For behind-the-meter rooftop photovoltaic systems, the energy flow becomes two-way. When there is more local generation than consumption, electricity is exported to the grid, allowing for net metering. However, electricity networks traditionally are not designed to deal with two-way energy transfer, which may introduce technical issues. An over-voltage issue may come out as the electricity flows from these PV households back to the network. There are solutions to manage the over-voltage issue, such as regulating PV inverter power factor, new voltage and energy control equipment at electricity distributor level, re-conductor the electricity wires, demand side management, etc. There are often limitations and costs related to these solutions.

  High generation during the middle of the day reduces the net generation demand, but higher peak net demand as the sun goes down can require rapid ramping of utility generating stations, producing a load profile called the duck curve.
- Implications for Electricity Bill Management and Energy Investment: There is no silver bullet in electricity or energy demand and bill management, because customers (sites) have different specific situations, e.g. different comfort/convenience needs, different electricity tariffs, or different usage patterns. Electricity tariff may have a few elements, such as daily access and metering charge, energy charge (based

on kWh, MWh) or peak demand charge (e.g. a price for the highest 30min energy consumption in a month). PV is a promising option for reducing energy charges when electricity prices are reasonably high and continuously increasing, such as in Australia and Germany. However, for sites with peak demand charge in place, PV may be less attractive if peak demands mostly occur in the late afternoon to early evening, for example in residential communities. Overall, energy investment is largely an economic decision and it is better to make investment decisions based on systematic evaluation of options in operational improvement, energy efficiency, onsite generation and energy storage.

## Photovoltaic Effect

The photovoltaic effect is the creation of voltage and electric current in a material upon exposure to light. It is a physical and chemical phenomenon.

The photovoltaic effect is closely related to the photoelectric effect. In either case, light is absorbed, causing excitation of an electron or other charge carrier to a higher-energy state. The main distinction is that the term photoelectric effect is now usually used when the electron is ejected out of the material (usually into a vacuum) and photovoltaic effect used when the excited charge carrier is still contained within the material. In either case, an electric potential (or voltage) is produced by the separation of charges, and the light has to have a sufficient energy to overcome the potential barrier for excitation. The physical essence of the difference is usually that photoelectric emission separates the charges by ballistic conduction and photovoltaic emission separates them by diffusion, but some "hot carrier" photovoltaic device concepts blur this distinction.

The first demonstration of the photovoltaic effect, by Edmond Becquerel in 1839, used an electrochemical cell. The first solar cell, consisting of a layer of selenium covered with a thin film of gold, was experimented by Charles Fritts in 1884, but it had a very poor efficiency. However, the most familiar form of the photovoltaic effect uses solid-state devices, mainly in photodiodes. When sunlight or other sufficiently energetic light is incident upon the photodiode, the electrons present in the valence band absorb energy and, being excited, jump to the conduction band and become free. These excited electrons diffuse, and some reach the rectifying junction (usually a diode p-n junction) where they are accelerated into the p-type semiconductor material by the built-in potential (Galvani potential). This generates an electromotive force and an electrical current, and thus some of the light energy is converted into electric energy. The photovoltaic effect can also occur when two photons are absorbed simultaneously in a process called two-photon photovoltaic effect.

The photovoltaic effect was first observed by French physicist A. E. Becquerel in 1839. He explained his discovery in Comptes rendus de l'Académie des sciences, "the production of an electric current when two plates of platinum or gold immersed in an acid, neutral, or alkaline solution are exposed in an uneven way to solar radiation."

Besides the direct excitation of free electrons, a photovoltaic effect can also arise simply due to the heating caused by absorption of the light. The heating leads to increased temperature of the semiconductor material, which is accompanied by temperature gradients. These thermal gradients in turn may generate a voltage through the Seebeck effect. Whether direct excitation or thermal effects dominate the photovoltaic effect will depend on many material parameters.

In most photovoltaic applications the radiation is sunlight, and the devices are called solar cells. In the case of a semiconductor p-n (diode) junction solar cell, illuminating the material creates an electric current because excited electrons and the remaining holes are swept in different directions by the built-in electric field of the depletion region.

## Photovoltaic System

A photovoltaic system, also PV system or solar power system, is a power system designed to supply usable solar power by means of photovoltaics. It consists of an arrangement of several components, including solar panels to absorb and convert sunlight into electricity, a solar inverter to convert the output from direct to alternating current, as well as mounting, cabling, and other electrical accessories to set up a working system. It may also use a solar tracking system to improve the system's overall performance and include an integrated battery solution, as prices for storage devices are expected to decline. Strictly speaking, a solar array only encompasses the ensemble of solar panels, the visible part of the PV system, and does not include all the other hardware, often summarized as balance of system (BOS). As PV systems convert light directly into electricity, they are not to be confused with other solar technologies, such as concentrated solar power or solar thermal, used for heating and cooling.

PV systems range from small, rooftop-mounted or building-integrated systems with capacities from a few to several tens of kilowatts, to large utility-scale power stations of hundreds of megawatts. Nowadays, most PV systems are grid-connected, while off-grid or stand-alone systems account for a small portion of the market.

Operating silently and without any moving parts or environmental emissions, PV systems have developed from being niche market applications into a mature technology used for mainstream electricity generation. A rooftop system recoups the invested energy for its manufacturing and installation within 0.7 to 2 years and produces about 95 percent of net clean renewable energy over a 30-year service lifetime.

Due to the growth of photovoltaics, prices for PV systems have rapidly declined since their introduction. However, they vary by market and the size of the system. In 2014, prices for residential 5-kilowatt systems in the United States were around $3.29 per watt, while in the highly penetrated German market, prices for rooftop systems of up to 100 kW declined to €1.24 per watt. Nowadays, solar PV modules account for less than half of the system's overall cost, leaving the rest to the remaining BOS-components and to soft costs, which include customer acquisition, permitting, inspection and interconnection, installation labor and financing costs.

## Modern System

A photovoltaic system converts the sun's radiation, in the form of light, into usable electricity. It comprises the solar array and the balance of system components. PV systems can be categorized by various aspects, such as, grid-connected vs. stand alone systems, building-integrated vs. rack-mounted systems, residential vs. utility systems, distributed vs. centralized systems, rooftop vs. ground-mounted systems, tracking vs. fixed-tilt systems, and new constructed vs. retrofitted systems. Other distinctions may include, systems with microinverters vs. central inverter, systems using crystalline silicon vs. Thin film technology, and systems with modules from Chinese vs. European and U.S.-manufacturers.

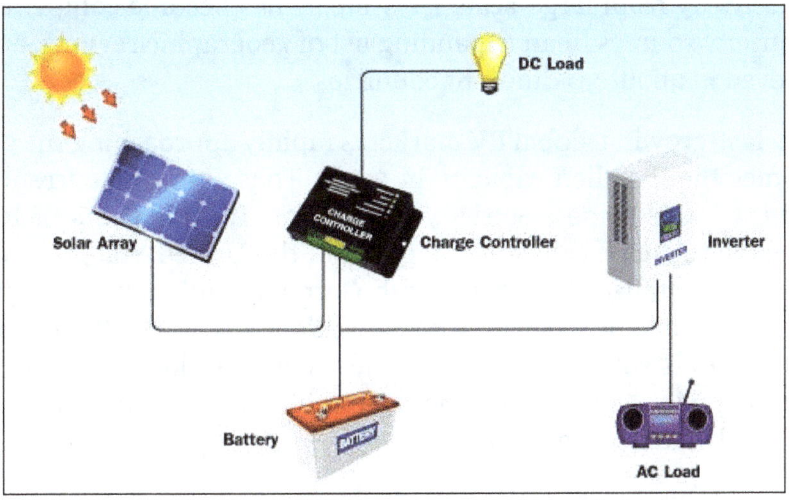


Diagram of the possible components of a photovoltaic system.

About 99 percent of all European and 90 percent of all U.S. solar power systems are connected to the electrical grid, while off-grid systems are somewhat more common in Australia and South Korea. PV systems rarely use battery storage. This may change, as government incentives for distributed energy storage are implemented and investments in storage solutions gradually become economically viable for small systems. A typical residential solar array is rack-mounted on the roof, rather than integrated into the roof or facade of the building, which is significantly more expensive. Utility-scale solar power stations are ground-mounted, with fixed tilted solar panels rather than using expensive tracking devices. Crystalline silicon is the predominant material used in 90 percent of worldwide produced solar modules, while its rival Thin film has lost market-share. About 70 percent of all solar cells and modules are produced in China and Taiwan, only 5 percent by European and US-manufacturers. The installed capacity for both, small rooftop systems and large solar power stations is growing rapidly and in equal parts, although there is a notable trend towards utility-scale systems, as the focus on new installations is shifting away from Europe to sunnier regions, such as the Sunbelt in the U.S., which are less opposed to ground-mounted solar farms and cost-effectiveness is more emphasized by investors.

Driven by advances in technology and increases in manufacturing scale and sophistication, the cost of photovoltaics is declining continuously. There are several million PV systems distributed all over the world, mostly in Europe, with 1.4 million systems in Germany alone – as well as North America with 440,000 systems in the United States, The energy conversion efficiency of a conventional solar module increased from 15 to 20 percent since 2004 and a PV system recoups the energy needed for its manufacture in about 2 years. In exceptionally irradiated locations, or when Thin film technology is used, the so-called energy payback time decreases to one year or less. Net metering and financial incentives, such as preferential feed-in tariffs for solar-generated electricity, have also greatly supported installations of PV systems in many countries. The levelised cost of electricity from large-scale PV systems has become competitive with conventional electricity sources in an expanding list of geographic regions, and grid parity has been achieved in about 30 different countries.

As of 2015, the fast-growing global PV market is rapidly approaching the 200 GW mark – about 40 times the installed capacity in 2006. These systems currently contribute about 1 percent to worldwide electricity generation. Top installers of PV systems in terms of capacity are currently China, Japan and the United States, while half of the world's capacity is installed in Europe, with Germany and Italy supplying 7% to 8% of their respective domestic electricity consumption with solar PV. The International Energy Agency expects solar power to become the world's largest source of electricity by 2050, with solar photovoltaics and concentrated solar thermal contributing 16% and 11% to the global demand, respectively.

## Grid-connection

A grid connected system is connected to a larger independent grid (typically the public electricity grid) and feeds energy directly into the grid. This energy may be shared by a residential or commercial building before or after the revenue measurement point, depending on whether the credited energy production is calculated independently of the customer's energy consumption (feed-in tariff) or only on the difference of energy (net metering). These systems vary in size from residential (2–10 $kW_p$) to solar power stations (up to 10s of $MW_p$). This is a form of decentralized electricity generation. Feeding electricity into the grid requires the transformation of DC into AC by a special, synchronising grid-tie inverter. In kilowatt-sized installations the DC side system voltage is as high as permitted (typically 1000 V except US residential 600 V) to limit ohmic losses. Most modules (60 or 72 crystalline silicon cells) generate 160 W to 300 W at 36 volts. It is sometimes necessary or desirable to connect the modules partially in parallel rather than all in series. An individual set of modules connected in series is known as a 'string'.

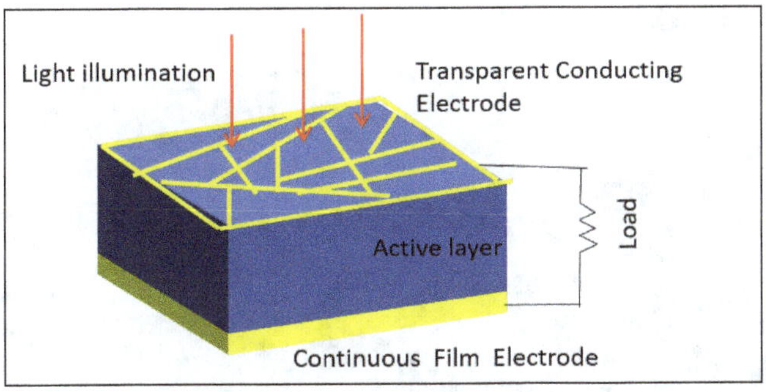


Schematics of a typical residential PV system.

## Scale of System

Photovoltaic systems are generally categorized into three distinct market segments: residential rooftop, commercial rooftop, and ground-mount utility-scale systems. Their capacities range from a few kilowatts to hundreds of megawatts. A typical residential system is around 10 kilowatts and mounted on a sloped roof, while commercial systems may reach a megawatt-scale and are generally installed on low-slope or even flat roofs. Although rooftop mounted systems are small and have a higher cost per watt than large utility-scale installations, they account for the largest share in the market. There is, however, a growing trend towards bigger utility-scale power plants, especially in the "sunbelt" region of the planet.

## Utility-scale

Perovo Solar Park in Ukraine.

Large utility-scale solar parks or farms are power stations and capable of providing an energy supply to large numbers of consumers. Generated electricity is fed into the transmission grid powered by central generation plants (grid-connected or grid-tied plant), or combined with one, or many, domestic electricity generators to feed into a small electrical grid (hybrid plant). In rare cases generated electricity is stored or used directly by island/standalone plant. PV systems are generally designed in order to ensure the highest energy yield for a given investment. Some large photovoltaic power stations such as Solar Star, Waldpolenz Solar Park and Topaz Solar Farm cover tens or hundreds of hectares and have power outputs up to hundreds of megawatts.

## Rooftop, Mobile and Portable

Rooftop system.

A small PV system is capable of providing enough AC electricity to power a single home, or an isolated device in the form of AC or DC electric. Military and civilian Earth observation satellites, street lights, construction and traffic signs, electric cars, solar-powered tents, and electric aircraft may contain integrated photovoltaic systems to provide a primary or auxiliary power source in the form of AC or DC power, depending on the design and power demands. In 2013, rooftop systems accounted for 60 percent of worldwide installations. However, there is a trend away from rooftop and towards utility-scale PV systems, as the focus of new PV installations is also shifting from Europe to countries in the sunbelt region of the planet where opposition to ground-mounted solar farms is less accentuated. Portable and mobile PV systems provide electrical power independent of utility connections, for "off the grid" operation. Such systems are so commonly used on recreational vehicles and boats that there are retailers specializing in these applications and products specifically targeted to them. Since recreational vehicles (RV) normally carry batteries and operate lighting and other systems on nominally 12-volt DC power, RV systems normally operate in a voltage range that can charge 12-volt batteries directly, so addition of a PV system requires only panels, a charge controller, and wiring. Solar systems on recreation vehicles are usually constrained in wattage by the physical size of the RV's roof space.

## Building-integrated

In urban and suburban areas, photovoltaic arrays are often used on rooftops to supplement power use; often the building will have a connection to the power grid, in which case the energy produced by the PV array can be sold back to the utility in some sort of net metering agreement. Some utilities, use the rooftops of commercial customers and telephone poles to support their use of PV panels. Solar trees are arrays that, as the name implies, mimic the look of trees, provide shade, and at night can function as street lights.

BAPV wall.

## Performance

Uncertainties in revenue over time relate mostly to the evaluation of the solar resource and to the performance of the system itself. In the best of cases, uncertainties are typically 4% for year-to-year climate variability, 5% for solar resource estimation (in a horizontal plane), 3% for estimation of irradiation in the plane of the array, 3% for power rating of modules, 2% for losses due to dirt and soiling, 1.5% for losses due to snow, and 5% for other sources of error. Identifying and reacting to manageable losses is critical for revenue and O&M efficiency. Monitoring of array performance may be part of contractual agreements between the array owner, the builder, and the utility purchasing the energy produced. A method to create "synthetic days" using readily available weather data and verification using the Open Solar Outdoors Test Field make it possible to predict photovoltaic systems performance with high degrees of accuracy. This method can be used to then determine loss mechanisms on a local scale - such as those from snow or the effects of surface coatings (e.g. hydrophobic or hydrophilic) on soiling or snow losses. (Although in heavy snow environments with severe ground interference can result in annual losses from snow of 30%.) Access to the Internet has allowed a further improvement in energy monitoring and communication. Dedicated systems are available from a number of vendors. For solar PV systems that use microinverters (panel-level DC to AC conversion), module power data is automatically provided. Some systems allow setting performance alerts that trigger phone/email/text warnings when limits are reached. These solutions provide data for the system owner and the installer. Installers are able to remotely monitor multiple installations, and see at-a-glance the status of their entire installed base.

## Components

A photovoltaic system for residential, commercial, or industrial energy supply consists of the solar array and a number of components often summarized as the balance of system (BOS). This term is synonymous with "Balance of plant" q.v. BOS-components include power-conditioning equipment and structures for mounting, typically one or more DC to AC power converters, also known as inverters, an energy storage device, a racking system that supports the solar array, electrical wiring and interconnections, and mounting for other components.

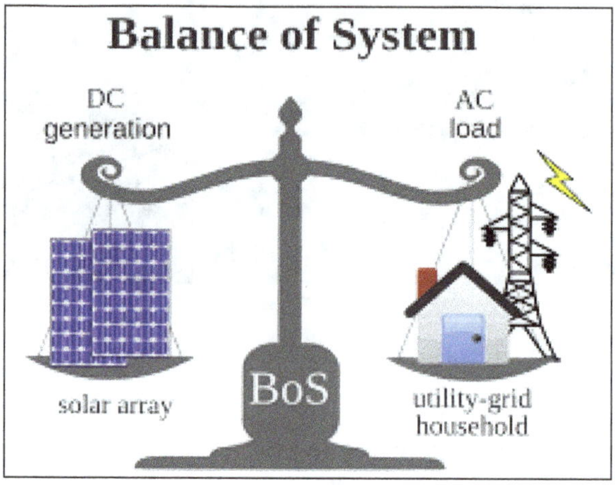


The balance of system components of a PV system (BOS) balance the power-generating subsystem of the solar array (left side) with the power-using side of the AC-household devices and the utility grid (right side).

Optionally, a balance of system may include any or all of the following: renewable energy credit revenue-grade meter, maximum power point tracker (MPPT), battery system and charger, GPS solar tracker, energy management software, solar irradiance sensors, anemometer, or task-specific accessories designed to meet specialized requirements for a system owner. In addition, a CPV system requires optical lenses or mirrors and sometimes a cooling system.

## References


- Solar-energy, science: britannica.com, Retrieved 13 April, 2020
- Pros-and-cons-solar-energy-053015: investopedia.com, Retrieved 18 June, 2020
- Ameri, Tayebeh; Dennler, Gilles; Lungenschmied, Christoph; Brabec, Christoph (2009). "Organic tandem solar cells: A review". Energy & Environmental Science. 2 (4): 348. doi:10.1039/B817952B. Retrieved 2019-05-20
- Solar-cell, technology: britannica.com, Retrieved 27 July, 2020
- Picault, D; Raison, B.; Bacha, S.; de la Casa, J.; Aguilera, J. (2010). "Forecasting photovoltaic array power production subject to mismatch losses" (PDF). Solar Energy. 84 (7): 1301–1309. Bibcode:2010SoEn...84.1301P. doi:10.1016/j.solener.2010.04.009. Retrieved 5 March 2013
- Solar-energy-10-major-application-of-solar-energy-explained, energy-28197: yourarticlelibrary.com, Retrieved 16 January, 2020
- Palz, Wolfgang (2013). Solar Power for the World: What You Wanted to Know about Photovoltaics. CRC Press. pp. 131–. ISBN 978-981-4411-87-5

# Hydropower 4

- Watermill
- Water Turbine
- Pumped-storage Hydroelectricity
- Small Hydro
- Micro Hydro
- Pico Hydro
- Marine Current Power
- Tidal Stream Generator

The power that is derived from the energy of falling or fast-running water is referred to as hydropower. Small hydro, micro hydro, pico hydro, marine current power, tidal stream generator, etc. are a few of its concepts. This chapter delves into the subject of hydropower for a thorough understanding of it.

Hydropower or hydroelectricity refers to the conversion of energy from flowing water into electricity. It is considered a renewable energy source because the water cycle is constantly renewed by the sun.

Historically, one of the first uses of hydro power was for mechanical milling, such as grinding grains. Today, modern hydro plants produce electricity using turbines and generators, where mechanical energy is created when moving water spins rotors on a turbine. This turbine is connected to an electromagnetic generator, which produce electricity when the turbine spins.

Hydro plant facilities can be categorized into three sizes: large (>30 MW), small (100 kW - 30 MW), or micro (<100 kW).

There are three main types of hydro plants:

- Impoundment facilities are the most common technology which uses a dam to create a large reservoir of water. Electricity is made when water passes through turbines in the dam.
- Pumped storage facilities are similar but have a second reservoir below the dam. Water can be pumped from the lower reservoir to the upper reservoir, storing energy for use at a later time.

- Run-of-river facilities rely more on natural water flow rates, diverting just a portion of river water through turbines, sometimes without the use of a dam or reservoirs. Since run-of-river hydro is subject to natural water variability, it is more intermittent than dammed hydro.

Hydropower is the largest contributor of all renewable energy sources and accounts for 6.7% of worldwide electricity production. Further growth of this mature technology may be possible, though many countries have already developed cost-effective sites.

Hydropower is an abundant, low cost source of power (where applicable), despite high upfront buidling costs. It is also a flexible and reliable source of electricity compared to other renewable options, as it may be stored for use at a later time. Dammed reservoirs can also help with flood control, be a reliable water supply, and may be used for recreational purposes.

However, there are many concerns with hydropower, particularly large dam facilities. Damming a river has a significant impact on the regional ecosystem, by flooding upstream landscapes, distrupting habitats for wildlife, blocking fish passages, and often displacing local communities. In addition, dam failures can be catastrophic, further disrupting landscapes and claiming the lives of those living downstream.

Finally, hydroplants are not completely free of greenhouse gas emissions. As with most forms of energy, carbon dioxide emissions occur during construction, particularly as a result of the large quantities of cement used, and loss of vegetation in flooded areas creates methane, another greenhouse gas, as it matter decays underwater.

## Generating Methods

### Conventional (Dams)

Generators at Hoover Dam.

Most hydroelectric power comes from the potential energy of dammed water driving a water turbine and generator. The power extracted from the water depends on the volume and on the difference in height between the source and the water's outflow. This height difference is called the head. A large pipe (the "penstock") delivers water from the reservoir to the turbine.

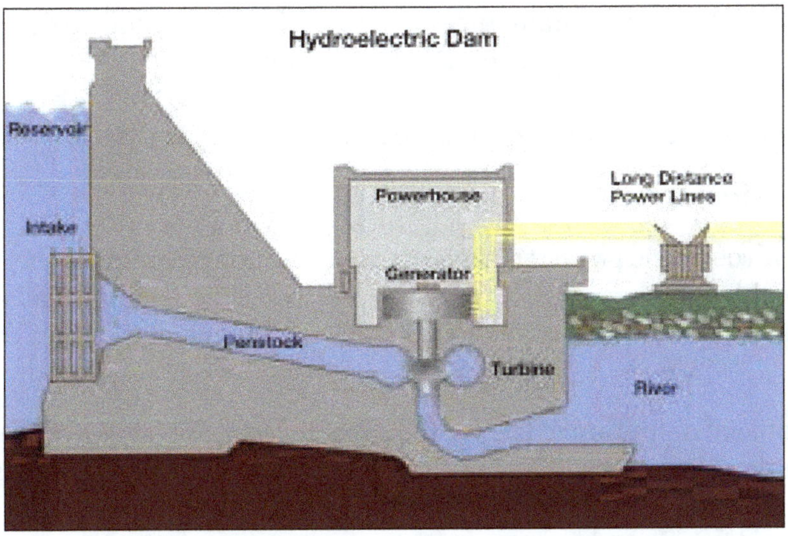


Cross section of a conventional hydroelectric dam.

## Pumped-storage

This method produces electricity to supply high peak demands by moving water between reservoirs at different elevations. At times of low electrical demand, the excess generation capacity is used to pump water into the higher reservoir. When the demand becomes greater, water is released back into the lower reservoir through a turbine. Pumped-storage schemes currently provide the most commercially important means of large-scale grid energy storage and improve the daily capacity factor of the generation system. Pumped storage is not an energy source, and appears as a negative number in listings.

## Run-of-the-river

Run-of-the-river hydroelectric stations are those with small or no reservoir capacity, so that only the water coming from upstream is available for generation at that moment, and any oversupply must pass unused. A constant supply of water from a lake or existing reservoir upstream is a significant advantage in choosing sites for run-of-the-river. In the United States, run of the river hydropower could potentially provide 60,000 megawatts (80,000,000 hp) (about 13.7% of total use in 2011 if continuously available).

## Tide

A tidal power station makes use of the daily rise and fall of ocean water due to tides; such sources are highly predictable, and if conditions permit construction of reservoirs, can also be dispatchable to generate power during high demand periods. Less common types of hydro schemes use water's kinetic energy or undammed sources such as undershot water wheels. Tidal power is viable in a relatively small number of locations around the world. In Great Britain, there are eight sites that could be developed, which have the potential to generate 20% of the electricity used in 2012.

## Sizes, Types and Capacities of Hydroelectric Facilities

### Large Facilities

Large-scale hydroelectric power stations are more commonly seen as the largest power producing facilities in the world, with some hydroelectric facilities capable of generating more than double the installed capacities of the current largest nuclear power stations.

Although no official definition exists for the capacity range of large hydroelectric power stations, facilities from over a few hundred megawatts are generally considered large hydroelectric facilities.

Panoramic view of the Itaipu Dam, with the spillways (closed at the time of the photo) on the left.

### Small

Small hydro is the development of hydroelectric power on a scale serving a small community or industrial plant. The definition of a small hydro project varies but a generating capacity of up to 10 megawatts (MW) is generally accepted as the upper limit of what can be termed small hydro. This may be stretched to 25 MW and 30 MW in Canada and the United States. Small-scale hydroelectricity production grew by 29% from 2005 to 2008, raising the total world small-hydro capacity to 85 GW. Over 70% of this was in China (65 GW), followed by Japan (3.5 GW), the United States (3 GW), and India (2 GW).

Small hydro stations may be connected to conventional electrical distribution networks as a source of low-cost renewable energy. Alternatively, small hydro projects may be built in isolated areas that would be uneconomic to serve from a network, or in areas where there is no national electrical distribution network. Since small hydro projects usually have minimal reservoirs and civil construction work, they are seen as having a relatively low environmental impact compared to large hydro. This decreased environmental impact depends strongly on the balance between stream flow and power production.

A micro-hydro facility.

Pico hydroelectricity.

## Micro

Micro hydro is a term used for hydroelectric power installations that typically produce up to 100 kW of power. These installations can provide power to an isolated home or small community, or are sometimes connected to electric power networks. There are many of these installations around the world, particularly in developing nations as they can provide an economical source of energy without purchase of fuel. Micro hydro systems complement photovoltaic solar energy systems because in many areas, water flow, and thus available hydro power, is highest in the winter when solar energy is at a minimum.

## Pico

Pico hydro is a term used for hydroelectric power generation of under 5 kW. It is useful in small, remote communities that require only a small amount of electricity. For example, to power one or two fluorescent light bulbs and a TV or radio for a few homes. Even smaller turbines of 200-300W may power a single home in a developing country with a drop of only 1 m (3 ft). A Pico-hydro setup is typically run-of-the-river, meaning that dams are not used, but rather pipes divert some of the flow, drop this down a gradient, and through the turbine before returning it to the stream.

## Underground

An underground power station is generally used at large facilities and makes use of a large natural height difference between two waterways, such as a waterfall or mountain lake. A tunnel is constructed to take water from the high reservoir to the generating hall built in a cavern near the lowest point of the water tunnel and a horizontal tailrace taking water away to the lower outlet waterway.

Measurement of the tailrace and forebay rates.

## Calculating Available Power

A simple formula for approximating electric power production at a hydroelectric station is:

$$P = -\eta\,(\dot{m} g\,\Delta h) = -\eta\,((\rho \dot{V})\,g\,\Delta h)$$

where:

- $P$ is power (in watts).
- $\eta$ ("eta") is the coefficient of efficiency (a unitless, scalar coefficient, ranging from 0 for completely inefficient to 1 for completely efficient).
- $\rho$ ("rho") is the density of water (~1000 kg/m$^3$).
- $\dot{V}$ is the volumetric flow rate (in m$^3$/s).
- $\dot{m}$ is the mass flow rate (in kg/s).
- $\Delta h$ ("Delta h") is the change in height (in meters).
- $g$ is acceleration due to gravity (9.8 m/s$^2$).

Efficiency is often higher (that is, closer to 1) with larger and more modern turbines. Annual electric energy production depends on the available water supply. In some installations, the water flow rate can vary by a factor of 10:1 over the course of a year.

## Advantages

### Flexibility

Hydropower is a flexible source of electricity since stations can be ramped up and down very quickly to adapt to changing energy demands. Hydro turbines have a start-up time of the order of a few minutes. It takes around 60 to 90 seconds to bring a unit from cold start-up to full load; this is much shorter than for gas turbines or steam plants. Power generation can also be decreased quickly when there is a surplus power generation. Hence the limited capacity of hydropower units is not generally used to produce base power except for vacating the flood pool or meeting downstream needs. Instead, it can serve as backup for non-hydro generators.

The Ffestiniog Power Station can generate 360 MW of electricity within 60 seconds of the demand arising.

## Low Cost/High Value Power

The major advantage of conventional hydroelectric dams with reservoirs is their ability to store water at low cost for dispatch later as high value clean electricity. The average cost of electricity from a hydro station larger than 10 megawatts is 3 to 5 U.S. cents per kilowatt-hour. When used as peak power to meet demand, hydroelectricity has a higher value than base power and a much higher value compared to intermittent energy sources.

Hydroelectric stations have long economic lives, with some plants still in service after 50–100 years. Operating labor cost is also usually low, as plants are automated and have few personnel on site during normal operation.

Where a dam serves multiple purposes, a hydroelectric station may be added with relatively low construction cost, providing a useful revenue stream to offset the costs of dam operation. It has been calculated that the sale of electricity from the Three Gorges Dam will cover the construction costs after 5 to 8 years of full generation. However, some data shows that in most countries large hydropower dams will be too costly and take too long to build to deliver a positive risk adjusted return, unless appropriate risk management measures are put in place.

## Suitability for Industrial Applications

While many hydroelectric projects supply public electricity networks, some are created to serve specific industrial enterprises. Dedicated hydroelectric projects are often built to provide the substantial amounts of electricity needed for aluminium electrolytic plants, for example. The Grand Coulee Dam switched to support Alcoa aluminium in Bellingham, Washington, United States for American World War II airplanes before it was allowed to provide irrigation and power to citizens (in addition to aluminium power) after the war. In Suriname, the Brokopondo Reservoir was constructed to provide electricity for the Alcoa aluminium industry. New Zealand's Manapouri Power Station was constructed to supply electricity to the aluminium smelter at Tiwai Point.

## Reduced $CO_2$ Emissions

Since hydroelectric dams do not use fuel, power generation does not produce carbon dioxide. While carbon dioxide is initially produced during construction of the project, and some methane is given off annually by reservoirs, hydro generally has the lowest lifecycle greenhouse gas emissions for power generation. Compared to fossil fuels generating an equivalent amount of electricity, hydro displaced three billion tonnes of $CO_2$ emissions in 2011. According to a comparative study by the Paul Scherrer Institute and the University of Stuttgart, hydroelectricity in Europe produces the least amount of greenhouse gases and externality of any energy source. Coming in second place was wind, third was nuclear energy, and fourth was solar photovoltaic. The low greenhouse gas impact of hydroelectricity is found especially in temperate climates. Greater greenhouse gas emission impacts are found in the tropical regions because the reservoirs of power stations in tropical regions produce a larger amount of methane than those in temperate areas.

Like other non-fossil fuel sources, hydropower also has no emissions of sulfur dioxide, nitrogen oxides, or other particulates.

## Other uses of the Reservoir

Reservoirs created by hydroelectric schemes often provide facilities for water sports, and become tourist attractions themselves. In some countries, aquaculture in reservoirs is common. Multi-use dams installed for irrigation support agriculture with a relatively constant water supply. Large hydro dams can control floods, which would otherwise affect people living downstream of the project.

## Disadvantages

## Ecosystem Damage and Loss of Land

Merowe Dam in Sudan: Hydroelectric power stations that use dams submerge large areas of land due to the requirement of a reservoir. These changes to land color or albedo, alongside certain projects that concurrently submerge rainforests, can in these specific cases result in the global warming impact, or equivalent life-cycle greenhouse gases of hydroelectricity projects, to potentially exceed that of coal power stations.

Large reservoirs associated with traditional hydroelectric power stations result in submersion of extensive areas upstream of the dams, sometimes destroying biologically rich and productive lowland and riverine valley forests, marshland and grasslands. Damming interrupts the flow of rivers and can harm local ecosystems, and building large dams and reservoirs often involves displacing people and wildlife. The loss of land is often exacerbated by habitat fragmentation of surrounding areas caused by the reservoir.

Hydroelectric projects can be disruptive to surrounding aquatic ecosystems both upstream and downstream of the plant site. Generation of hydroelectric power changes the downstream river environment. Water exiting a turbine usually contains very little suspended sediment, which can lead to scouring of river beds and loss of riverbanks. Since turbine gates are often opened intermittently, rapid or even daily fluctuations in river flow are observed.

## Water Loss by Evaporation

Hydroelectric plants in the U.S. consumed between 5.39 to 68.14 cubic metres per megawatt-hour (1,425 to 18,000 US gallons per megawatt-hour) of electricity generated, through evaporation losses in the reservoir. The median loss was 17.00 $m^3$/MWh (4,491 US gal/MWh), which is higher than the loss for generation technologies that use cooling towers, including concentrating solar power at 3.27 $m^3$/MWh (865 US gal/MWh) for CSP trough and 2.98 $m^3$/MWh (786 US gal/MWh) for CSP tower, coal at 2.60 $m^3$/MWh (687 US gal/MWh), nuclear at 2.54 $m^3$/MWh (672 US gal/MWh), and natural gas at 0.75 $m^3$/MWh (198 US gal/MWh). Where there are multiple uses of reservoirs such as water supply, recreation, and flood control, all reservoir evaporation is attributed to power production.

## Siltation and Flow Shortage

When water flows it has the ability to transport particles heavier than itself downstream. This has a negative effect on dams and subsequently their power stations, particularly those on rivers or within catchment areas with high siltation. Siltation can fill a reservoir and reduce its capacity to control floods along with causing additional horizontal pressure on the upstream portion of the dam. Eventually, some reservoirs can become full of sediment and useless or over-top during a flood and fail.

Changes in the amount of river flow will correlate with the amount of energy produced by a dam. Lower river flows will reduce the amount of live storage in a reservoir therefore reducing the amount of water that can be used for hydroelectricity. The result of diminished river flow can be power shortages in areas that depend heavily on hydroelectric power. The risk of flow shortage may increase as a result of climate change. One study from the Colorado River in the United States suggest that modest climate changes, such as an increase in temperature in 2 degree Celsius resulting in a 10% decline in precipitation, might reduce river run-off by up to 40%. Brazil in particular is vulnerable due to its heavy reliance on hydroelectricity, as increasing temperatures, lower water flow and alterations in the rainfall regime, could reduce total energy production by 7% annually by the end of the century.

## Methane Emissions from Reservoirs

The Hoover Dam in the United States is a large conventional dammed-hydro facility, with an installed capacity of 2,080 MW.

Lower positive impacts are found in the tropical regions, as it has been noted that the reservoirs of power plants in tropical regions produce substantial amounts of methane. This is due to plant material in flooded areas decaying in an anaerobic environment and forming methane, a greenhouse gas. According to the World Commission on Dams report, where the reservoir is large compared to the generating capacity (less than 100 watts per square metre of surface area) and no clearing of the forests in the area was undertaken prior to impoundment of the reservoir, greenhouse gas emissions from the reservoir may be higher than those of a conventional oil-fired thermal generation plant.

In boreal reservoirs of Canada and Northern Europe, however, greenhouse gas emissions are typically only 2% to 8% of any kind of conventional fossil-fuel thermal generation. A new class of underwater logging operation that targets drowned forests can mitigate the effect of forest decay.

## Relocation

Another disadvantage of hydroelectric dams is the need to relocate the people living where the reservoirs are planned. In 2000, the World Commission on Dams estimated that dams had physically displaced 40-80 million people worldwide.

## Failure Risks

Because large conventional dammed-hydro facilities hold back large volumes of water, a failure due to poor construction, natural disasters or sabotage can be catastrophic to downriver settlements and infrastructure.

During Typhoon Nina in 1975 Banqiao Dam failed in Southern China when more than a year's worth of rain fell within 24 hours. The resulting flood resulted in the deaths of 26,000 people, and another 145,000 from epidemics. Millions were left homeless.

The creation of a dam in a geologically inappropriate location may cause disasters such as 1963 disaster at Vajont Dam in Italy, where almost 2,000 people died.

The Malpasset Dam failure in Fréjus on the French Riviera (Côte d'Azur), southern France, collapsed on December 2, 1959, killing 423 people in the resulting flood.

Smaller dams and micro hydro facilities create less risk, but can form continuing hazards even after being decommissioned. For example, the small earthen embankment Kelly Barnes Dam failed in 1977, twenty years after its power station was decommissioned, causing 39 deaths.

## Comparison and Interactions with other Methods of Power Generation

Hydroelectricity eliminates the flue gas emissions from fossil fuel combustion, including pollutants such as sulfur dioxide, nitric oxide, carbon monoxide, dust, and mercury in the coal. Hydroelectricity also avoids the hazards of coal mining and the indirect health effects of coal emissions.

## Nuclear Power

Compared to nuclear power, hydroelectricity construction requires altering large areas of the environment while a nuclear power station has a small footprint, and hydro-powerstation failures have caused tens of thousands of more deaths than any nuclear station failure. The creation of Garrison Dam, for example, required Native American land to create Lake Sakakawea, which has a shoreline of 2,120 kilometres (1,320 mi), and caused the inhabitants to sell 94% of their arable land for $7.5 million in 1949.

However, nuclear power is relatively inflexible; although nuclear power can reduce its output reasonably quickly. Since the cost of nuclear power is dominated by its high infrastructure costs, the cost per unit energy goes up significantly with low production. Because of this, nuclear power is mostly used for baseload. By way of contrast, hydroelectricity can supply peak power at much lower cost. Hydroelectricity is thus often used to complement nuclear or other sources for load following. Country examples were they are paired in a close to 50/50 share include the electric grid in Switzerland, the Electricity sector in Sweden and to a lesser extent, Ukraine and the Electricity sector in Finland.

### Wind Power

Wind power goes through predictable variation by season, but is intermittent on a daily basis. Maximum wind generation has little relationship to peak daily electricity consumption, the wind may peak at night when power isn't needed or be still during the day when electrical demand is highest. Occasionally weather patterns can result in low wind for days or weeks at a time, a hydroelectric reservoir capable of storing weeks of output is useful to balance generation on the grid. Peak wind power can be offset by minimum hydropower and minimum wind can be offset with maximum hydropower. In this way the easily regulated character of hydroelectricity is used to compensate for the intermittent nature of wind power. Conversely, in some cases wind power can be used to spare water for later use in dry seasons.

In areas that do not have hydropower, pumped storage serves a similar role, but at a much higher cost and 20% lower efficiency. An example of this is Norway's trading with Sweden, Denmark, the Netherlands and possibly Germany or the UK in the future. Norway is 98% hydropower, while its flatland neighbors are installing wind power.

## Watermill

A watermill is a structure that uses a water wheel to drive a mechanical process such as grinding flour, lumber or textile production, or metal shaping. A water mill that generates electricity is usually called a hydroelectric power plant.

The first documented use of a water mill was in the first century BC and the technology spread quite quickly across the world.

Commercial mills were in use in Roman Britain and by the time of the Doomsday Book in the late 11th Century there were more than 6,000 water mills in England.

By the 16th Century waterpower was the most important source of motive power in Britain and Europe. The number of mills probably peaked at more than 20,000 mills by the 19th Century.

The Norman Conquest introduced the feudal system to the country and 'soke rights' forced everyone to have their corn milled at the mill owned by their Manorial Lord which stayed in use until as late as the 19th Century.

Although the "Right of Soke" was never written in law, it was commonly practised in Britain – and throughout Europe – until the 16th century, even after the country had moved away from feudalism.

With agricultural developments at the beginning of the Industrial Revolution, people began to move away from agriculture into other industries.

Rural water mills began to close down to be replaced by the large, industrial, port-based steam-powered mill and by the end of the 19th Century almost all rural watermills had ceased commercial production.

## How it Works?

Water mills use the flow of water to turn a large water wheel. A shaft connected to the wheel axle is then used to transmit the power from the water through a system of gears and cogs to work machinery, such as a millstone to grind corn.

Watermills are usually built beside streams or rivers to use them as a water supply. Very often these supplies were improved by the provision of mill races and weirs to help overcome the problems of different seasonal water levels. Many of the weirs seen on rivers today were originally built to help control water levels for watermills.

Usually power from the waterwheel was transferred to the inside of the mill by a shaft extending from the waterwheel axle which connected to a large gear wheel called the Pit Wheel (because half of it lies below the floor level of the mill, in a pit). Power was then turned through 90 degrees by bevel gears to turn a large vertically mounted shaft leading to the upper floor of the water mill.

At the lower level, this shaft drove a large Spur Wheel, which then provided power to turn the millstones. At the top of the shaft another large wheel, the Crown Wheel, was placed to drive other machinery in the mill, such as sack hoists.

Millwrights were highly skilled craftsmen who could work with wood and iron to build the complex working machinery of the mill.

## Types of Watermills

There were many types of watermill. You will find the main types listed below:

- Gristmills, or corn mills, grind grains into flour. These were undoubtedly the most common kind of mill.
- Fulling or walk mills were used for a finishing process on cloth.
- Sawmills cut timber into lumber.
- Bark Mills stripped bark from trees or ground it to powder for use in tanneries.
- Spoke mills turned lumber into spokes for carriage wheels.
- Cotton mills were usually powered by a water wheel at the beginning of the industrial revolution.
- Bobbin Mills made wooden bobbins for the cotton and other textile industries.
- Carpet mills for making rugs were sometimes water-powered.
- Textile mills for weaving cloth were sometimes water-powered.
- Powder mills for making gunpowder - black powder or smokeless powder were usually water-powered.
- Blast Furnaces, finery forges, and tinplate works were water powered. Furnaces and Forges were sometimes called iron mills.
- Blade mills were used for sharpening newly made blades.
- Slitting mills were used for slitting bars of iron into rods, which were then made into nails.
- Rolling mills shaped metal by passing it between rollers.
- Lead was usually smelted in smeltmills prior to the introduction of the cupola (a reverberatory furnace).
- Paper mills used water not only for motive power, but also required it in large quantities in the manufacturing process.

# Water Turbine

A water turbine is a rotary machine that converts kinetic energy and potential energy of water into mechanical work.

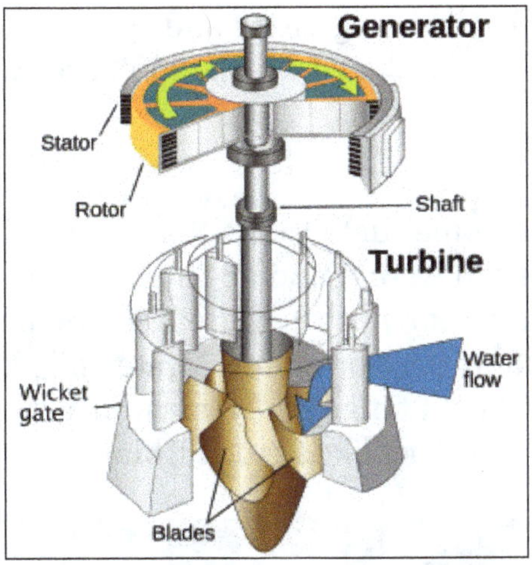


Kaplan turbine and electrical generator cut-away view.

Water turbines were developed in the 19th century and were widely used for industrial power prior to electrical grids. Now they are mostly used for electric power generation. Water turbines are mostly found in dams to generate electric power from water potential energy.

The construction of a Ganz water Turbo Generator.

Water wheels have been used for hundreds of years for industrial power. Their main shortcoming is size, which limits the flow rate and head that can be harnessed. The migration from water wheels to modern turbines took about one hundred years. Development occurred during the Industrial revolution, using scientific principles and methods. They also made extensive use of new materials and manufacturing methods developed at the time.

## Swirl

The main difference between early water turbines and water wheels is a swirl component of the water which passes energy to a spinning rotor. This additional component of motion allowed the turbine to be smaller than a water wheel of the same power. They could process more water by spinning faster and could harness much greater heads. (Later, impulse turbines were developed which didn't use swirl.)

## Timeline

The earliest known water turbines date to the Roman Empire. Two helix-turbine mill sites of almost identical design were found at Chemtou and Testour, modern-day Tunisia, dating to the late 3rd or early 4th century AD. The horizontal water wheel with angled blades was installed at the bottom of a water-filled, circular shaft. The water from the mill-race entered the pit tangentially, creating a swirling water column which made the fully submerged wheel act like a true turbine.

A Francis turbine runner, rated at nearly one million hp (750 MW).

A propeller-type runner rated 28,000 hp (21 MW).

Fausto Veranzio in his book described a vertical axis mill with a rotor similar to that of a Francis turbine.

Johann Segner developed a reactive water turbine (Segner wheel) in the mid-18th century in Kingdom of Hungary. It had a horizontal axis and was a precursor to modern water turbines. It is a very simple machine that is still produced today for use in small hydro sites. Segner worked with Euler on some of the early mathematical theories of turbine design. In the 18th century, a Dr. Robert Barker invented a similar reaction hydraulic turbine that became popular as a lecture-hall demonstration. The only known surviving example of this type of engine used in power production, dating from 1851, is found at Hacienda Buena Vista in Ponce, Puerto Rico.

In 1820, Jean-Victor Poncelet developed an inward-flow turbine.

In 1826, Benoît Fourneyron developed an outward-flow turbine. This was an efficient machine (~80%) that sent water through a runner with blades curved in one dimension. The stationary outlet also had curved guides.

In 1844, Uriah A. Boyden developed an outward flow turbine that improved on the performance of the Fourneyron turbine. Its runner shape was similar to that of a Francis turbine.

In 1849, James B. Francis improved the inward flow reaction turbine to over 90% efficiency. He also conducted sophisticated tests and developed engineering methods for water turbine design. The Francis turbine, named for him, is the first modern water turbine. It is still the most widely used water turbine in the world today. The Francis turbine is also called a radial flow turbine, since water flows from the outer circumference towards the centre of runner.

Inward flow water turbines have a better mechanical arrangement and all modern reaction water turbines are of this design. As the water swirls inward, it accelerates, and transfers energy to the runner. Water pressure decreases to atmospheric, or in some cases subatmospheric, as the water passes through the turbine blades and loses energy.

Around 1890, the modern fluid bearing was invented, now universally used to support heavy water turbine spindles. As of 2002, fluid bearings appear to have a mean time between failures of more than 1300 years.

Around 1913, Viktor Kaplan created the Kaplan turbine, a propeller-type machine. It was an evolution of the Francis turbine and revolutionized the ability to develop low-head hydro sites.

## New Concept

All common water machines until the late 19th century (including water wheels) were basically reaction machines; water pressure head acted on the machine and produced work. A reaction turbine needs to fully contain the water during energy transfer.

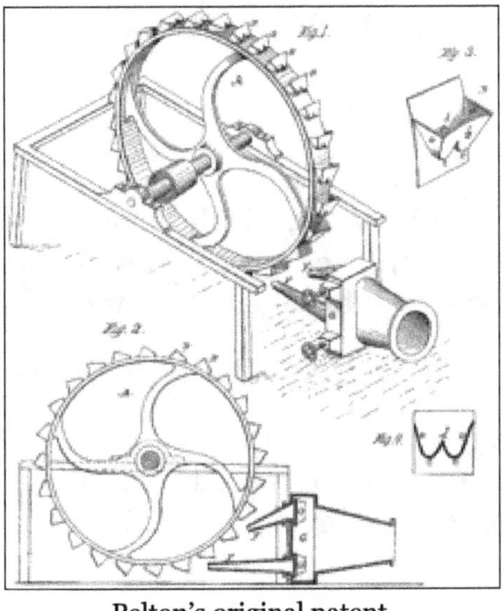

Pelton's original patent.

In 1866, California millwright Samuel Knight invented a machine that took the impulse system to a new level. Inspired by the high pressure jet systems used in hydraulic mining in the gold fields, Knight developed a bucketed wheel which captured the energy of a free jet, which had converted a high head (hundreds of vertical feet in a pipe or penstock) of water to kinetic energy. This is called an impulse or tangential turbine. The water's velocity, roughly twice the velocity of the bucket periphery, does a U-turn in the bucket and drops out of the runner at low velocity.

In 1879, Lester Pelton, experimenting with a Knight Wheel, developed a Pelton wheel (double bucket design), which exhausted the water to the side, eliminating some energy loss of the Knight wheel which exhausted some water back against the center of the wheel. In about 1895, William Doble improved on Pelton's half-cylindrical bucket form with an elliptical bucket that included a cut in it to allow the jet a cleaner bucket entry. This is the modern form of the Pelton turbine which today achieves up to 92% efficiency. Pelton had been quite an effective promoter of his design and although Doble took over the Pelton company he did not change the name to Doble because it had brand name recognition.

Turgo and cross-flow turbines were later impulse designs.

## Theory of Operation

Flowing water is directed on to the blades of a turbine runner, creating a force on the blades. Since the runner is spinning, the force acts through a distance (force acting through a distance is the definition of work). In this way, energy is transferred from the water flow to the turbine.

Water turbines are divided into two groups: reaction turbines and impulse turbines.

The precise shape of water turbine blades is a function of the supply pressure of water, and the type of impeller selected.

## Reaction Turbines

Reaction turbines are acted on by water, which changes pressure as it moves through the turbine and gives up its energy. They must be encased to contain the water pressure (or suction), or they must be fully submerged in the water flow.

Newton's third law describes the transfer of energy for reaction turbines.

Most water turbines in use are reaction turbines and are used in low (<30 m or 100 ft) and medium (30–300 m or 100–1,000 ft) head applications. In reaction turbine pressure drop occurs in both fixed and moving blades. It is largely used in dam and large power plants

## Impulse Turbines

Impulse turbines change the velocity of a water jet. The jet pushes on the turbine's curved blades which changes the direction of the flow. The resulting change in momentum (impulse) causes a force on the turbine blades. Since the turbine is spinning, the force acts through a distance (work) and the diverted water flow is left with diminished energy. An impulse turbine is one in which the pressure of the fluid flowing over the rotor blades is constant and all the work output is due to the change in kinetic energy of the fluid.

Prior to hitting the turbine blades, the water's pressure (potential energy) is converted to kinetic energy by a nozzle and focused on the turbine. No pressure change occurs at the turbine blades, and the turbine doesn't require a housing for operation.

Newton's second law describes the transfer of energy for impulse turbines.

Impulse turbines are often used in very high (>300m/1000 ft) head applications.

## Power

The power available in a stream of water is;

$$P = P = \eta \cdot \rho \cdot g \cdot h \cdot \dot{q}$$

where:

- $P =$ power (J/s or watts).
- $\eta =$ turbine efficiency.
- $\rho =$ density of water (kg/m$^3$).
- $g =$ acceleration of gravity (9.81 m/s$^2$).
- $h =$ head (m). For still water, this is the difference in height between the inlet and outlet surfaces. Moving water has an additional component added to account for the kinetic energy of the flow. The total head equals the pressure head plus velocity head.
- $\dot{q}$ = flow rate (m$^3$/s).

## Pumped-storage Hydroelectricity

Some water turbines are designed for pumped-storage hydroelectricity. They can reverse flow and operate as a pump to fill a high reservoir during off-peak electrical hours, and then revert to a water turbine for power generation during peak electrical demand. This type of turbine is usually a Deriaz or Francis turbine in design.

This type of system is used in El Hierro, one of the Canary Islands: "When wind production exceeds demand, excess energy will pump water from a lower reservoir at the bottom of a volcanic cone to an upper reservoir at the top of the volcano 700 meters above sea level. The lower reservoir stores 150,000 cubic meters of water. The stored water acts as a battery. The maximum storage capacity is 270 MWh. When demand rises and there is not enough wind power, the water will be released to four hydroelectric turbines with a total capacity of 11 MW."

## Efficiency

Large modern water turbines operate at mechanical efficiencies greater than 90%.

## Types of Water Turbines

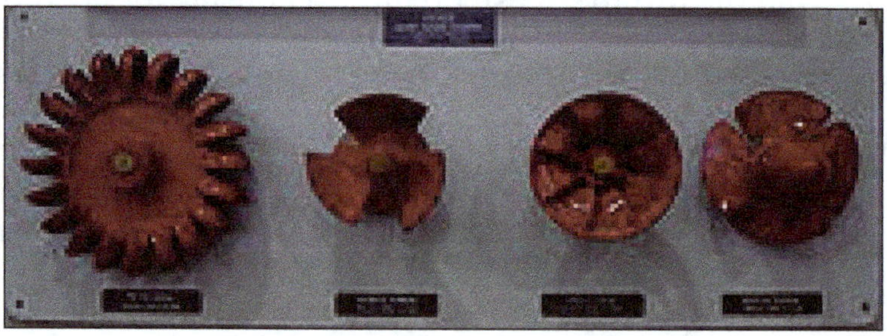

Various types of water turbine runners. From left to right: Pelton wheel two types of Francis turbine and Kaplan turbine.

Reaction turbines:

- VLH turbine.
- Francis turbine.
- Kaplan turbine.
- Tyson turbine.
- Deriaz turbine.
- Gorlov helical turbine.

Impulse turbine:

- Water wheel.
- Pelton wheel.
- Turgo turbine.
- Cross-flow turbine (also known as the Bánki-Michell turbine, or Ossberger turbine).
- Jonval turbine.
- Reverse overshot water-wheel.
- Screw turbine.
- Barkh Turbine.

## Design and Application

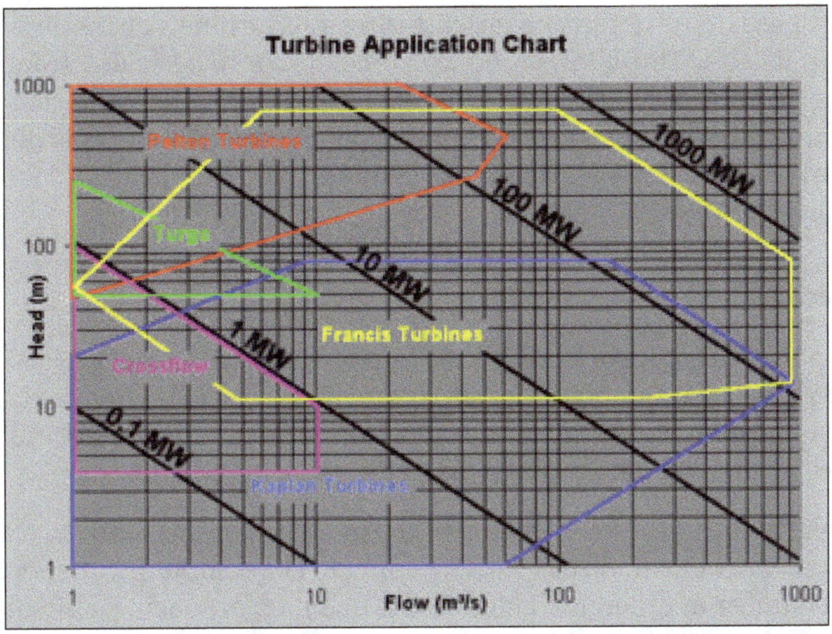


Turbine selection is based on the available water head, and less so on the available flow rate. In general, impulse turbines are used for high head sites, and reaction turbines are used for low head sites. Kaplan turbines with adjustable blade pitch are well-adapted to wide ranges of flow or head conditions, since their peak efficiency can be achieved over a wide range of flow conditions.

Small turbines (mostly under 10 MW) may have horizontal shafts, and even fairly large bulb-type turbines up to 100 MW or so may be horizontal. Very large Francis and Kaplan machines usually have vertical shafts because this makes best use of the available head, and makes installation of a generator more economical. Pelton wheels may be either vertical or horizontal shaft machines because the size of the machine is so much less than the available head. Some impulse turbines use multiple jets per runner to balance shaft thrust. This also allows for the use of a smaller turbine runner, which can decrease costs and mechanical losses.

## Typical Range of Heads

| Water wheel | $0.2 < H < 4$ (H = head in m) |
|---|---|
| Screw turbine | $1 < H < 10$ |
| VLH turbine | $1.5 < H < 4.5$ |
| Kaplan turbine | $2 < H < 70$ |
| Francis turbine | $10 < H < 300$ |
| Pelton wheel | $80 < H < 1600$ |
| Turgo turbine | $50 < H < 250$ |

## Specific Speed

The specific speed $n_s$ of a turbine characterizes the turbine's shape in a way that is not related to its size. This allows a new turbine design to be scaled from an existing design of known performance. The specific speed is also the main criteria for matching a specific hydro site with the correct turbine type. The specific speed is the speed with which the turbine turns for a particular discharge Q, with unit head and thereby is able to produce unit power.

## Affinity Laws

Affinity laws allow the output of a turbine to be predicted based on model tests. A miniature replica of a proposed design, about one foot (0.3 m) in diameter, can be tested and the laboratory measurements applied to the final application with high confidence. Affinity laws are derived by requiring similitude between the test model and the application.

Flow through the turbine is controlled either by a large valve or by wicket gates arranged around the outside of the turbine runner. Differential head and flow can be plotted for a number of different values of gate opening, producing a hill diagram used to show the efficiency of the turbine at varying conditions.

## Runaway Speed

The runaway speed of a water turbine is its speed at full flow, and no shaft load. The turbine will be designed to survive the mechanical forces of this speed. The manufacturer will supply the runaway speed rating.

## Control Systems

Different designs of governors have been used since the mid-18th century to control the speeds of the water turbines. A variety of flyball systems, or first-generation governors, were used during the first 100 years of water turbine speed controls. In early flyball systems, the flyball component countered by a spring acted directly to the valve of the turbine or the wicket gate to control the amount of water that enters the turbines. Newer systems with mechanical governors started around 1880. An early mechanical governor is a servomechanism that comprises a series of gears that use the turbine's speed to drive the flyball and turbine's power to drive the control mechanism. The mechanical governors were continued to be enhanced in power amplification through the use of gears and the dynamic behavior. By 1930, the mechanical governors had many parameters that could be set on the feedback system for precise controls. In the later part of the twentieth century, electronic governors and digital systems started to replace the mechanical governors. In the electronic governors, also known as second-generation governors, the flyball was replaced by rotational speed sensor but the controls were still done through analog systems. In the modern systems, also known as third-generation governors, the controls are performed digitally by algorithms that are programmed to the computer of the governor.

## Turbine Blade Materials

Given that the turbine blades in a water turbine are constantly exposed to water and dynamic forces, they need to have high corrosion resistance and strength. The most common material used in overlays on carbon steel runners in water turbines are austenitic steel alloys that have 17% to 20% chromium to increase stability of the film which improves aqueous corrosion resistance. The chromium content in these steel alloys exceed the minimum of 12% chromium required to exhibit some atmospheric corrosion resistance. Having a higher chromium concentration in the steel alloys allows for a much longer lifespan of the turbine blades. Currently, the blades are made of martensitic stainless steels which have high strength compared to austenitic stainless steels by a factor of 2. Besides corrosion resistance and strength as the criteria for material selection, weld-ability and density of the turbine blade. Greater weld-ability allows for easier repair of the turbine blades. This also allows for higher weld quality which results in a better repair. Selecting a material with low density is important to achieve higher efficiency because the lighter blades rotate more easily. The most common material used in Kaplan Turbine blades are stainless steel alloys (SS). The martensitic stainless steel alloys have high strength, thinner sections than standard carbon steel, and reduced mass that enhances the hydrodynamic flow conditions and efficiency of the water turbine. The SS(13Cr-4Ni) has been shown to have improved erosion resistance at all angles of attack through the process of laser hardening. It is important to minimize erosion in order to maintain high efficiencies because erosion negatively impacts the hydraulic profile of the blades which reduces the relative ease to rotate.

## Maintenance

Turbines are designed to run for decades with very little maintenance of the main elements; overhaul intervals are on the order of several years. Maintenance of the runners and parts exposed to water include removal, inspection, and repair of worn parts.

A Francis turbine at the end of its life showing pitting corrosion, fatigue cracking and a catastrophic failure. Earlier repair jobs that used stainless steel weld rods are visible.

Normal wear and tear includes pitting corrosion from cavitation, fatigue cracking, and abrasion from suspended solids in the water. Steel elements are repaired by welding, usually with stainless steel rods. Damaged areas are cut or ground out, then welded back up to their original or an improved profile. Old turbine runners may have a significant amount of stainless steel added this way by the end of their lifetime. Elaborate welding procedures may be used to achieve the highest quality repairs.

Other elements requiring inspection and repair during overhauls include bearings, packing box and shaft sleeves, servomotors, cooling systems for the bearings and generator coils, seal rings, wicket gate linkage elements and all surfaces.

## Environmental Impact

Walchensee Hydroelectric Power Station in Bavaria, Germany, has been in operation since 1924.

Water turbines are generally considered a clean power producer, as the turbine causes essentially no change to the water. They use a renewable energy source and are designed to operate for decades. They produce significant amounts of the world's electrical supply.

Historically there have also been negative consequences, mostly associated with the dams normally required for power production. Dams alter the natural ecology of rivers, potentially killing fish, stopping migrations, and disrupting peoples' livelihoods. For example, Native American tribes in the Pacific Northwest had livelihoods built around salmon fishing, but aggressive dam-building destroyed their way of life. Dams also cause less obvious, but potentially serious consequences, including increased evaporation of water (especially in arid regions), buildup of silt behind the dam, and changes to water temperature and flow patterns. In the United States, it is now illegal to block the migration of fish, for example the white sturgeon in North America, so fish ladders must be provided by dam builders.

## Pumped-storage Hydroelectricity

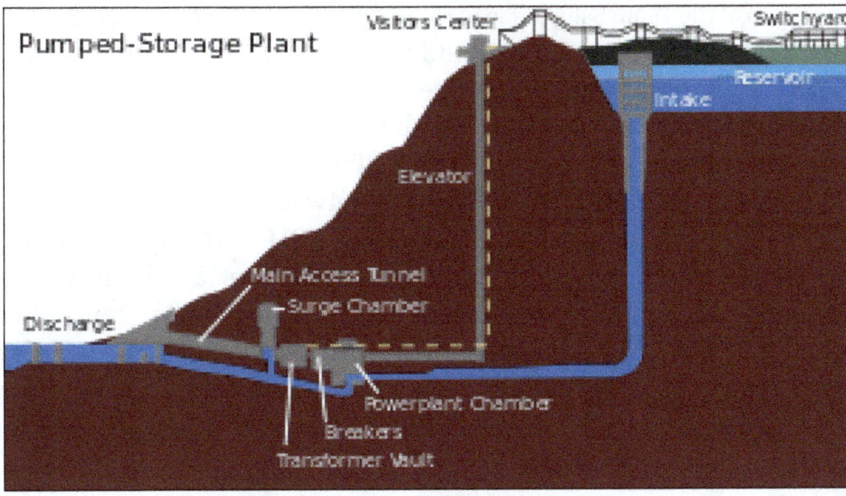


Diagram of the TVA pumped storage facility.

Pumped-storage hydroelectricity (PSH), or pumped hydroelectric energy storage (PHES), is a type of hydroelectric energy storage used by electric power systems for load balancing. The method stores energy in the form of gravitational potential energy of water, pumped from a lower elevation reservoir to a higher elevation. Low-cost surplus off-peak electric power is typically used to run the pumps. During periods of high electrical demand, the stored water is released through turbines to produce electric power. Although the losses of the pumping process makes the plant a net consumer of energy overall, the system increases revenue by selling more electricity during periods of peak demand, when electricity prices are highest.

Pumped-storage hydroelectricity allows energy from intermittent sources (such as solar, wind) and other renewables, or excess electricity from continuous base-load sources (such as coal or nuclear) to be saved for periods of higher demand. The reservoirs used with pumped storage are quite small when compared to conventional hydroelectric dams of similar power capacity, and generating periods are often less than half a day.

Shaded-relief topo map of the Taum Sauk pumped storage plant. Pumped storage is the largest-capacity form of grid energy storage available, and, as of 2017, the United States Department of Energy Global Energy Storage Database reports that PSH accounts for over 95% of all active tracked storage installations worldwide, with a total installed nameplate capacity of over 184 GW, of which about 25 GW are in the United States. The round-trip energy efficiency of PSH varies between 70%–80%, with some sources claiming up to 87%. The main disadvantage of PSH is the specialist nature of the site required, needing both geographical height and water availability. Suitable sites are therefore likely to be in hilly or mountainous regions, and potentially in areas of outstanding natural beauty, and therefore there are also social and ecological issues to overcome. Many recently proposed projects, at least in the U.S., avoid highly sensitive or scenic areas, and some propose to take advantage of "brownfield" locations such as disused mines.

## Basic Principle

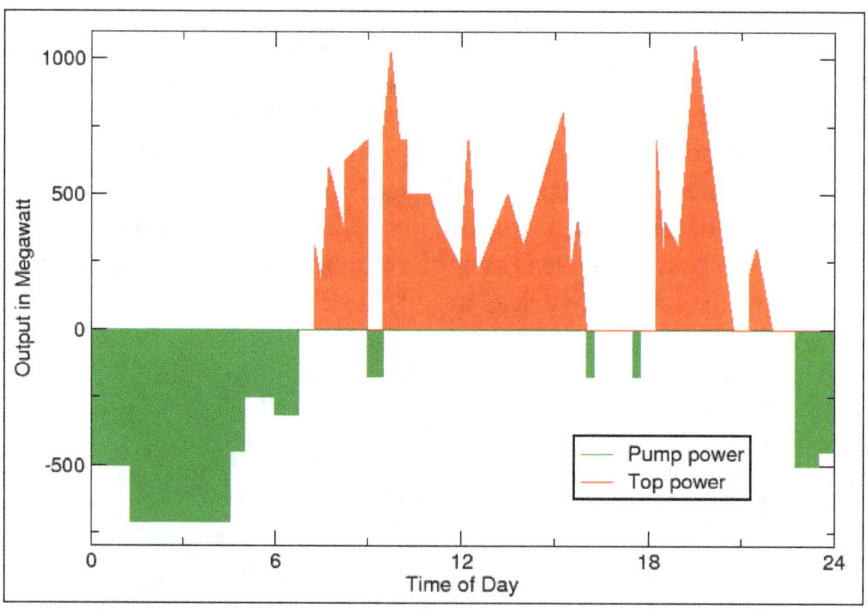


Power distribution, over a day, of a pumped-storage hydroelectricity facility. Green represents power consumed in pumping; red is power generated.

At times of low electrical demand, excess generation capacity is used to pump water into the upper reservoir. When there is higher demand, water is released back into the lower reservoir through a turbine, generating electricity. Reversible turbine/generator assemblies act as a combined pump and turbine generator unit (usually a Francis turbine design). In micro-PSH applications, a group of pumps and Pump As Turbine (PAT) could be implemented respectively for pumping and generating phases. The same pump could be used in both modes by changing rotational direction and speed: the operation point in pumping usually differs by operation point in PAT mode.

## Types: Natural or Man-made Reservoirs

In open-loop systems, pure pumped-storage plants store water in an upper reservoir with no natural inflows, while pump-back plants utilize a combination of pumped storage and conventional hydroelectric plants with an upper reservoir that is replenished in part by natural inflows from a stream or river. Plants that do not use pumped-storage are referred to as conventional hydroelectric plants; conventional hydroelectric plants that have significant storage capacity may be able to play a similar role in the electrical grid as pumped storage by deferring output until needed.

## Economic Efficiency

Taking into account evaporation losses from the exposed water surface and conversion losses, energy recovery of 70-80% or more can be achieved. This technique is currently the most cost-effective means of storing large amounts of electrical energy, but capital costs and the presence of appropriate geography are critical decision factors in selecting pumped-storage plant sites.

The relatively low energy density of pumped storage systems requires either large flows large differences in height between reservoirs. The only way to store a significant amount of energy is by having a large body of water located relatively near, but as high above as possible, a second body of water. In some places this occurs naturally, in others one or both bodies of water were man-made. Projects in which both reservoirs are artificial and in which no natural inflows are involved with either reservoir are referred to as "closed loop" systems.

These systems may be economical because they flatten out load variations on the power grid, permitting thermal power stations such as coal-fired plants and nuclear power plants that provide base-load electricity to continue operating at peak efficiency, while reducing the need for "peaking" power plants that use the same fuels as many base-load thermal plants, gas and oil, but have been designed for flexibility rather than maximal efficiency. Hence pumped storage systems are crucial when coordinating large groups of heterogeneous generators. Capital costs for pumped-storage plants are relatively high, although this is somewhat mitigated by their long service life of up to 75 years or more, which is three to five times longer than utility-scale batteries.

Along with energy management, pumped storage systems help control electrical network frequency and provide reserve generation. Thermal plants are much less able to respond to sudden changes in electrical demand, potentially causing frequency and voltage instability. Pumped storage plants, like other hydroelectric plants, can respond to load changes within seconds.

The upper reservoir (Llyn Stwlan) and dam of the Ffestiniog Pumped Storage Scheme. The lower power station has four water turbines which generate 360 MW of electricity within 60 seconds of the need arising.

The most important use for pumped storage has traditionally been to balance baseload powerplants, but may also be used to abate the fluctuating output of intermittent energy sources. Pumped storage provides a load at times of high electricity output and low electricity demand, enabling additional system peak capacity. In certain jurisdictions, electricity prices may be close to zero or occasionally negative on occasions that there is more electrical generation available than there is load available to absorb it; although at present this is rarely due to wind or solar power alone, increased wind and solar generation will increase the likelihood of such occurrences. It is particularly likely that pumped storage will become especially important as a balance for very large scale photovoltaic generation. Increased long distance transmission capacity combined with significant amounts of energy storage will be a crucial part of regulating any large-scale deployment of intermittent renewable power sources. The high non-firm renewable electricity penetration in some regions supplies 40% of annual output, but 60% may be reached before additional storage is necessary.

## Small-scale Facilities

While smaller scale pumped storage experiences an economy of scale penalty, there are small-scale installations of such technology, including a recent 13 MW project in Germany. Shell Energy has proposed a 5 MW project in Washington State. Some have proposed small pumped storage plants in buildings, although these are economically unfeasible given the economies of scale present. Also, a large volume of water is required for a meaningful storage capacity which is a difficult fit for an urban setting. Nevertheless, some authors defend its technological simplicity and secure provision of water as important externalities.

## Worldwide Use

Adam Beck Complex.

In 2009, world pumped storage generating capacity was 104 GW, while other sources claim 127 GW, which comprises the vast majority of all types of utility grade electric storage. The EU had 38.3 GW net capacity (36.8% of world capacity) out of a total of 140 GW of hydropower and representing 5% of total net electrical capacity in the EU. Japan had 25.5 GW net capacity (24.5% of world capacity).

In 2010 the United States had 21.5 GW of pumped storage generating capacity (20.6% of world capacity). PSH generated (net) -5.501 GWh of energy in 2010 in the US because more energy is consumed in pumping than is generated. Nameplate pumped storage capacity had grown to 21.6 GW by 2014, with pumped storage comprising 97% of grid-scale energy storage in the US. As of late 2014, there were 51 active project proposals with a total of 39 GW of new nameplate capacity across all stages of the FERC licensing process for new pumped storage hydroelectric plants in the US, but no new plants were currently under construction in the US at the time.

In June 2018 the Australian federal government announced that 14 sites had been identified in Tasmania for pumped storage hydro, with the potential of adding 4.8GW to the national grid if a second interconnector beneath Bass Strait was constructed.

## Pump-back Hydroelectric Dams

Conventional hydroelectric dams may also make use of pumped storage in a hybrid system that both generates power from water naturally flowing into the reservoir as well as storing water pumped back to the reservoir from below the dam. The Grand Coulee Dam in the US was expanded with a pump-back system in 1973. Existing dams may be repowered with reversing turbines thereby extending the length of time the plant can operate at capacity. Optionally a pump back powerhouse such as the Russell Dam may be added to a dam for increased generating capacity. Making use of an existing dam's upper reservoir and transmission system can expedite projects and reduce costs.

In January 2019, the State Grid Corporation of China announced plans to invest US$5.7 billion in five pumped hydro storage plants with a total 6GW capacity, to be located in Hebei, Jilin, Zhejiang, Shandong provinces, and in Xinjiang Autonomous Region. China is seeking to build 40GW of pumped hydro capacity installed by 2020.

## Potential Technologies

### Seawater

Pumped storage plants can operate with seawater, although there are additional challenges compared to using fresh water. Inaugurated in 1966, the 240 MW Rance tidal power station in France can partially work as a pumped-storage station. When high tides occur at off-peak hours, the turbines can be used to pump more seawater into the reservoir than the high tide would have naturally brought in. It is the only large-scale power plant of its kind.

In 1999, the 30 MW Yanbaru project in Okinawa was the first demonstration of seawater pumped storage. It has since been decommissioned. A 300 MW seawater-based Lanai Pumped Storage Project was considered for Lanai, Hawaii, and seawater-based projects have been proposed in Ireland. A pair of proposed projects in the Atacama Desert in northern Chile would use 600 MW of photovoltaic solar (Skies of Tarapacá) together with 300 MW of pumped storage (Mirror of Tarapacá) raising seawater 600 metres (2,000 ft) up a coastal cliff.

### Underground Reservoirs

The use of underground reservoirs has been investigated. Recent examples include the proposed Summit project in Norton, Ohio, the proposed Maysville project in Kentucky (underground limestone mine), and the Mount Hope project in New Jersey, which was to have used a former iron mine as the lower reservoir. The proposed energy storage at the Callio site in Pyhäjärvi (Finland) would utilize the deepest base metal mine in Europe, with 1,450 metres (4,760 ft) elevation difference. Several new underground pumped storage projects have been proposed. Cost-per-kilowatt estimates for these projects can be lower than for surface projects if they use existing underground mine space. There are limited opportunities involving suitable underground space, but the number of underground pumped storage opportunities may increase if abandoned coal mines prove suitable.

In Bendigo, Victoria, Australia, the Bendigo Sustainability Group has proposed the use of the old gold mines under Bendigo for Pumped Hydro Energy Storage. Bendigo has the greatest concentration of deep shaft hard rock mines anywhere in the world with over 5,000 shafts sunk under Bendigo in the second half of the 19th Century. The deepest shaft extends 1,406 metres vertically underground. A recent pre-feasibility study has shown the concept to be viable with a generation capacity of 30 MW and a run time of 6 hours using a water head of over 750 metres.

## Decentralised Systems

Small (or micro) applications for pumped-storage could be built on streams and within infrastructures, such as drinking water networks and artificial snow making infrastructures. In this regard, a storm-water basin has been concretely implemented as a cost-effective solution for a water reservoir in a micro pumped hydro energy storage. Such plants provide distributed energy storage and distributed flexible electricity production and can contribute to the decentralized integration of intermittent renewable energy technologies, such as wind power and solar power. Reservoirs that can be used for small pumped-storage hydropower plants could include natural or artificial lakes, reservoirs within other structures such as irrigation, or unused portions of mines or underground military installations. In Switzerland one study suggested that the total installed capacity of small pumped-storage hydropower plants in 2011 could be increased by 3 to 9 times by providing adequate policy instruments.

## Underwater Reservoirs

In March 2017 the research project StEnSea (Storing Energy at Sea) announced their successful completion of a four-week test of a pumped storage underwater reservoir. In this configuration a hollow sphere submerged and anchored at great depth acts as the lower reservoir, while the upper reservoir is the enclosing body of water. Electricity is created when water is let in via a reversible turbine integrated into the sphere. During off-peak hours the turbine changes direction and pumps the water out again, using "surplus" electricity from the grid. The quantity of power created when water is let in grows proportionally to the height of the column of water above the sphere, in other words: the deeper the sphere is located, the more potential energy it can store, which can be transformed into electric power. On the other hand, pumping the water back out at greater depths also uses up more power, since the turbine-turned-pump must act on the same entire column of water.

As such the energy storage capacity of the submerged reservoir is not governed by the gravitational energy in the traditional sense, but rather by the vertical pressure variation.

While StEnSea's test took place at a depth of 100 m in the fresh water Lake Constance, the technology is foreseen to be used in salt water at greater depths. Since the submerged reservoir needs only a connecting electrical cable, the depth at which it can be employed is limited only by the depth at which the turbine can function, currently limited to 700 m. The challenge of designing salt water pumped storage in this underwater configuration brings a range of advantages:

- No land area is required.
- No mechanical structure other than the electrical cable needs to span the distance of the potential energy difference.
- In the presence of sufficient seabed area multiple reservoirs can scale the storage capacity without limits.

- Should a reservoir collapse, the consequences would be limited apart from the loss of the reservoir itself.
- Evaporation from the upper reservoir has no effect on the energy conversion efficiency.
- Transmission of electricity between the reservoir and the grid can be established from a nearby offshore wind farm limiting transmission loss and obviating the need for onshore cabling permits.

A current commercial design featuring a sphere with an inner diameter of 30 m submerged to 700 m would correspond to a 20 MWh capacity which with a 5 MW turbine would lead to a 4-hour discharge time. An energy park with multiple such reservoirs would bring the storage cost to around a few eurocents per kWh with construction and equipment costs in the range €1,200-€1,400 per kW. To avoid excessive transmission cost and loss, the reservoirs should be placed off deep water coasts of densely populated areas, such as Norway, Spain, USA and Japan. With this limitation the concept would allow for worldwide electricity storage of close to 900 GWh.

For comparison, a traditional, gravity-based pumped storage capable of storing 20 MWh in a water reservoir the size of a 30 m sphere would need a hydraulic head of 519 m with the elevation spanned by a pressurized water pipe requiring typically a hill or mountain for support.

### Home Use

Using a pumped-storage system of cisterns and small generators, pico hydro may also be effective for "closed loop" home energy generation systems.

## Small Hydro

Small hydro is the development of hydroelectric power on a scale suitable for local community and industry, or to contribute to distributed generation in a regional electricity grid. The definition of a "small hydro" project varies based on different consideration of what should constitute bottom and top limits from country to country, but most common range, although not binding, in capacity of "small hydro" extend between 1 and 15 megawatts (MW), and can be further subdivide by scale into "mini" (<1MW), "micro" (<100 kW), "pico" (<10 kW). In contrast many hydroelectric projects are of enormous size, such as the generating plant at the Three Gorges Dam at 22,500 megawatts or the vast multiple projects of the Tennessee Valley Authority. In India, hydro projects up to 25 MW station capacities have been categorized as Small Hydro Power (SHP) projects.

Small power plant.

An 1895 hydroelectric plant.

Small hydro projects may be built in isolated areas that would be uneconomic to serve from a national electricity grid, or in areas where a national grid does not exist.

The use of the term "small hydro" varies considerably around the world, the maximum limit is usually somewhere between 10 and 30 MW. While a minimum limit is not usually set, the US National Hydropower Association specifies a minimum limit of 5 MW. In California, hydroelectric generating stations with a maximum capacity of less than 30 MW are classified as small, and are eligible for inclusion in the state's renewable portfolio standard, while hydroelectric generating stations with a higher capacity are classified as large and are not considered renewable. The "small hydro" description may be stretched up to 50 MW in the United States, Canada and China. Small hydro can be further subdivided into mini hydro, usually defined as 100 to 1,000 kilowatts (kW), and micro hydro which is 5 to 100 kW. Micro hydro is usually the application of hydroelectric power sized for smaller communities, single families or small enterprise. The smallest installations are pico hydro, below 5 kW.

Since small hydro projects usually have correspondingly small civil construction work and little or no reservoir, they are seen as having a relatively low environmental impact compared to large hydro.

## Growth

According to a report by REN21, during 2008 small hydro installations grew by 28% over year 2005 to raise the total world small hydro capacity to 85 gigawatts (GW). Over 70% of this was in China (with 65 GW), followed by Japan (3.5 GW), the United States (3 GW) and India (2 GW). Global growth is expected to be 2.8% annually until the mid-2020s when capacity will be about 150 gigawatts.

China planned to electrify a further 10,000 villages between 2005 and 2010 under their China Village Electrification Program, including further investments in small hydro and photovoltaics. By 2010, China had 45,000 small hydro installations, especially in rural areas, producing 160 Twh annually. Over 50% of the world's potential small hydro power was found in Asia; however, the report noted, "It is possible in the future that more small hydropower potential might be identified both on the African and American continents".

In the mountains and rain forests of British Columbia, Canada there are a great many sites suitable for hydro development. However environmental concerns towards large reservoirs after the 1980s halted new dam construction. The solution to coping with increased demand was to offer contracts to independent power producers, who have built 100 run of the river projects under 50 MW. Power production without reservoirs varies dramatically, but older conventional dams retain or release water to average out production though the year. In 2014 these independent producers generated 18,000 GWh from 4,500 MW of capacity.

## Generation

Historic Ottenbach Small Hydro with original equipment of 1920, still runningfor guided visits.

Hydroelectric power is the generation of electric power from the movement of water. A hydroelectric facility requires a dependable flow of water and a reasonable height for the water to fall, called the head. In a typical installation, water is fed from a reservoir through a pipe into a turbine. The water flowing through the turbine causes an electrical generator to rotate, converting the motion into electrical energy.

Hongping Power station has a design typical for small hydro stations. Water comes from the mountain behind the station, through the black pipe.

Small hydro may developed by constructing new facilities or through re-development of existing dams whose primary purpose is flood control, or irrigation. Old hydro sites may be re-developed, sometimes salvaging substantial investment in the installation such as penstock pipe and turbines, or just re-using the water rights associated with an abandoned site. Either of these cost saving advantages can make the return on investment for a small hydro site well worth the use of existing sites.

## Project Design

Many companies offer standardized turbine generator packages in the approximate size range of 200 kW to 10 MW. These "water to wire" packages simplify the planning and development of the site since one vendor looks after most of the equipment supply. Because non-recurring engineering costs are minimized and development cost is spread over multiple units, the cost of such package systems is reduced. While synchronous generators capable of isolated plant operation are often used, small hydro plants connected to an electrical grid system can use economical induction generators to further reduce installation cost and simplify control and operation.

Small "run of the river" projects do not have a conventional dam with a reservoir, only a weir to form a headpond for diversion of inlet water to the turbine. Unused water simply flows over the weir and the headpond may only be capable of a single day's storage, not enough for dry summers or frozen winters when generation may come to a halt. A preferred scenario is to have the inlet in an existing lake.

Modular "micro hydrokinetic" systems have been developed for irrigation canals. "Irrigation districts across the U.S. have installed power plants at diversion points and in-canal drops, which are traditionally used for flow measurement, to stabilize upstream heads and to dissipate energy where there is significant elevation change throughout the canal system."

Countries like India and China have policies in favor of small hydro, and the regulatory process allows for building dams and reservoirs. In North America and Europe the regulatory process is too long and expensive to consider having a dam and a reservoir for a small project.

Small hydro projects usually have faster environmental and licensing procedures, and since the equipment is usually in serial production, standardized and simplified, and the civil works construction is also reduced, the projects may be developed very rapidly. The physically smaller size of equipment makes it easier to transport to remote areas without good road or rail access.

One measure of decreased environmental impact with lakes and reservoirs depends on the balance between stream flow and power production. Reducing water diversions helps the river's ecosystem, but reduces the hydro system's return on Investment (ROI). The hydro system design must strike a balance to maintain both the health of the stream and the economics.

# Micro Hydro

Micro hydro.

Micro hydro is a type of hydroelectric power that typically produces from 5 kW to 100 kW of electricity using the natural flow of water. Installations below 5 kW are called pico hydro. These installations can provide power to an isolated home or small community, or are sometimes connected to electric power networks, particularly where net metering is offered. There are many of these installations around the world, particularly in developing nations as they can provide an economical source of energy without the purchase of fuel. Micro hydro systems complement solar PV power systems because in many areas, water flow, and thus available hydro power, is highest in the winter when solar energy is at a minimum. Micro hydro is frequently accomplished with a pelton wheel for high head, low flow water supply. The installation is often just a small dammed pool, at the top of a waterfall, with several hundred feet of pipe leading to a small generator housing. In low head sites, generally water wheels and Archimedes screws are used.

## Construction

Construction details of a microhydro plant are site-specific. Sometimes an existing mill-pond or other artificial reservoir is available and can be adapted for power production. In general, microhydro systems are made up of a number of components. The most important include the intake where water is diverted from the natural stream, river, or perhaps a waterfall. An intake structure such as a catch box is required to screen out floating debris and fish, using a screen or array of bars to keep out large objects. In temperate climates, this structure must resist ice as well. The intake may have a gate to allow the system to be dewatered for inspection and maintenance.

The intake is then brought through a canal and then forebay. The forebay is used for sediment holding. At the bottom of the system the water is tunneled through a pipeline (penstock) to the powerhouse building containing a turbine. The penstock builds up pressure from the water that has traveled downwards. In mountainous areas, access to the route of the penstock may provide considerable challenges. If the water source and turbine are far apart, the construction of the penstock may be the largest part of the costs of construction. At the turbine, a controlling valve is installed to regulate the flow and the speed of the turbine. The turbine converts the flow and pressure of the water to mechanical energy; the water emerging from the turbine returns to the natural watercourse along a tailrace channel. The turbine turns a generator, which is then connected to electrical loads; this might be directly connected to the power system of a single building in very small installations, or may be connected to a community distribution system for several homes or buildings.

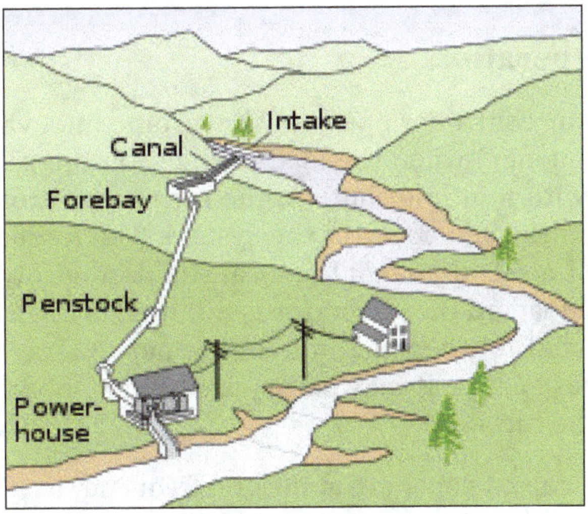


Typical microhydro setup.

Usually, microhydro installations do not have a dam and reservoir, like large hydroelectric plants have, relying on a minimal flow of water to be available year-round.

## Head and Flow Characteristics

Microhydro systems are typically set up in areas capable of producing up to 100 kilowatts of electricity. This can be enough to power a home or small business facility. This production range is calculated in terms of "head" and "flow". The higher each of these are, the more power available. Hydraulic head is the pressure measurement of water falling in a pipe expressed as a function of the vertical distance the water falls. This change in elevation is usually measured in feet or meters. A drop of at least 2 feet is required or the system may not be feasible. When quantifying head, both gross and net head must be considered. Gross head approximates power accessibility through the vertical distance measurement alone whereas net head subtracts pressure lost due to friction in piping from the gross head. "Flow" is the actual quantity of water falling from a site and is usually measured in gallons per minute, cubic feet per second, or liters per second. Low flow/high head installations in steep terrain have significant pipe costs. A long penstock starts with low pressure pipe at the top and progressively higher pressure pipe closer to the turbine in order to reduce pipe costs.

The available power, in kilowatts, from such a system can be calculated by the equation P=Q*H/k, where Q is the flow rate in gallons per minute, H is the static head, and k is a constant of 5,310 gal*ft/min*kW. For instance, for a system with a flow of 500 gallons per minute and a static head of 60 feet, the theoretical maximum power output is 5.65 kW. The system is prevented from 100% efficiency (from obtaining all 5.65 kW) due to the real world, such as: turbine efficiency, friction in pipe, and conversion from potential to kinetic energy. Turbine efficiency is generally between 50-80%, and pipe friction is accounted for using the Hazen–Williams equation.

## Regulation and Operation

Typically, an automatic controller operates the turbine inlet valve to maintain constant speed (and frequency) when the load changes on the generator. In a system connected to a grid with multiple sources, the turbine control ensures that power always flows out from the generator to the system. The frequency of the alternating current generated needs to match the local standard utility frequency. In some systems, if the useful load on the generator is not high enough, a load bank may be automatically connected to the generator to dissipate energy not required by the load; while this wastes energy, it may be required if it's not possible to control the water flow through the turbine.

An induction generator always operates at the grid frequency irrespective of its rotation speed; all that is necessary is to ensure that it is driven by the turbine faster than the synchronous speed so that it generates power rather than consuming it. Other types of generator can use a speed control systems for frequency matching.

With the availability of modern power electronics it is often easier to operate the generator at an arbitrary frequency and feed its output through an inverter which produces output at grid frequency. Power electronics now allow the use of permanent magnet alternators that produce wild AC to be stabilised. This approach allows low speed/low head water turbines to be competitive; they can run at the best speed for extraction of energy, and the power frequency is controlled by the electronics instead of the generator.

Very small installations (pico hydro), a few kilowatts or smaller, may generate direct current and charge batteries for peak use times.

## Turbine Types

Several types of water turbines can be used in micro hydro installations, selection depending on the head of water, the volume of flow, and such factors as availability of local maintenance and transport of equipment to the site. For hilly regions where a waterfall of 50 meters or more may be available, a Pelton wheel can be used. For low head installations, Francis or propeller-type turbines are used. Very low head installations of only a few meters may use propeller-type turbines in a pit, or water wheels and Archimedes screws. Small micro hydro installations may successfully use industrial centrifugal pumps, run in reverse as prime movers; while the efficiency may not be as high as a purpose-built runner, the relatively low cost makes the projects economically feasible.

In low-head installations, maintenance and mechanism costs can be relatively high. A low-head system moves larger amounts of water, and is more likely to encounter surface debris. For this reason a Banki turbine also called Ossberger turbine, a pressurized self-cleaning crossflow waterwheel, is often preferred for low-head micro hydro systems. Though less efficient, its simpler structure is less expensive than other low-head turbines of the same capacity. Since the water flows in, then out of it, it cleans itself and is less prone to jam with debris.

- Screw turbine (Reverse Archimedes' screw): Two low-head schemes in England, Settle Hydro and Torrs Hydro use an Archimedes' screw which is another debris-tolerant design. Efficiency 85%.
- Gorlov: The Gorlov helical turbine free stream or constrained flow with or without a dam,

Francis and propeller turbines:

- Kaplan turbine: Is a high flow, low head, propeller-type turbine. An alternative to the traditional kaplan turbine is a large diameter, slow turning, permanent magnet, sloped open flow VLH turbine with efficiencies of 90%.
- Water wheel: Advanced hydraulic water wheels and hydraulic wheel-part reaction turbine can have hydraulic efficiencies of 67% and 85% respectively. Overshot water wheel maximum efficiency (hydraulic efficiency) is 85%. Undershot water wheels can operate with very low head, but also have efficiencies below 30%.
- Gravitation water vortex power plant: Part of the river flow at a weir or natural water fall is diverted into a round basin with a central bottom exit that creates a vortex. A simple rotor (and connected generator) is moved by the kinetic energy. Efficiencies of 83% down to 64% at 1/3 part flow.

## Use

Microhydro systems are very flexible and can be deployed in a number of different environments. They are dependent on how much water flow the source (creek, river, stream) has and the velocity of the flow of water. Energy can be stored in battery banks at sites that are far from a facility or used in addition to a system that is directly connected so that in times of high demand there is additional reserve energy available. These systems can be designed to minimize community and environmental impact regularly caused by large dams or other mass hydroelectric generation sites.

## Potential for Rural Development

In relation to rural development, the simplicity and low relative cost of micro hydro systems open up new opportunities for some isolated communities in need of electricity. With only a small stream needed, remote areas can access lighting and communications for homes, medical clinics, schools, and other facilities. Microhydro can even run a certain level of machinery supporting small businesses. Regions along the Andes mountains and in Sri Lanka and China already have similar, active programs. One seemingly unexpected use of such systems in some areas is to keep young community members from moving into more urban regions in order to spur economic growth. Also, as the possibility of financial incentives for less carbon intensive processes grows, the future of microhydro systems may become more appealing.

Micro-hydro installations can also provide multiple uses. For instance, micro-hydro projects in rural Asia have incorporated agro-processing facilities such as rice mills – alongside standard electrification – into the project design.

## Advantages and Disadvantages

### Advantages

Microhydro power is generated through a process that utilizes the natural flow of water. This power is most commonly converted into electricity. With no direct emissions resulting from this conversion process, there are little to no harmful effects on the environment, if planned well, thus supplying power from a renewable source and in a sustainable manner. Microhydro is considered a "run-of-river" system meaning that water diverted from the stream or river is redirected back into the same watercourse. Adding to the potential economic benefits of microhydro is efficiency, reliability, and cost effectiveness.

### Disadvantages

Microhydro systems are limited mainly by characteristics of the site. The most direct limitation comes from small sources with minuscule flow. Likewise, flow can fluctuate seasonally in some areas. Lastly, though perhaps the foremost disadvantage is the distance from the power source to the site in need of energy. This distributional issue as well as the others are key when considering using a microhydro system.

# Pico Hydro

Pico hydro is a term used for hydroelectric power generation of under 5 kW. These generators have proven to be useful in small, remote communities that require only a small amount of electricity – for example, to power one or two fluorescent light bulbs and a TV or radio in 50 or so homes. Even smaller turbines of 200–300 W may power a single home in a developing country with a drop of only one meter. Pico-hydro setups typically are run-of-stream, meaning that a reservoir of water is not created, only a small weir is common, pipes divert some of the flow, drop this down a gradient, and through the turbine before being exhausted back to the stream.

A pico hydro system made by the Sustainable Vision project.

Like other hydroelectric and renewable source power generation, pollution and consumption of fossil fuels is reduced, though there is still typically an environmental cost to the manufacture of the generator and distribution methods.

## Small-scale DIY Hydroplants

With a growing DIY-community and an increasing interest in environmentally friendly "green energy", some hobbyists have endeavored to build their own hydroelectric plants from old water mills, from kits, or from scratch. The DIY-community has used abandoned water mills to mount a waterwheel and electrical generating components. This approach has also been popularized in the TV-series It's Not Easy Being Green. These are usually smaller turbines of ~5 kW or less. Through the internet, the community is now able to obtain plans to construct DIY-water turbines, and there is a growing trend toward building them for domestic requirements. The DIY-hydroelectric plants are now being used both in developed countries and in developing countries, to power residences and small businesses. Two examples of pico hydro power can be found in the towns of Kithamba and Thimba in the Central Province of Kenya. These produce 1.1 kW and 2.2 kW, respectively. Local residents were trained to maintain the hydro schemes. The pico hydro sites in Kenya won Ashden Awards for Sustainable Energy.

Using a pumped-storage system of cisterns and small generators, pico hydro may also be effective for "closed loop" home systems.

## Manufacturers

In Vietnam, several Chinese manufacturers have sold pico-powerplants at prices as low as $20–70 for a powerplant of 300–500 W. However, the devices sold are said to be low in quality and may damage connected equipment if connected improperly.

Sam Redfield of the Appropriate Infrastructure Development Group (AIDG) has developed a pico-hydro generator made from common PVC pipe and a modified Toyota alternator housed in a five gallon bucket. The generator was developed to provide power to communities without access to the electricity grid in developing countries. Envisioned as an energy source to charge cell phones, provide lighting and charge batteries, the generator is designed to be made by artisans with basic skills and can be built for less than $150.00. The Toyota alternator used in the generator is converted to a permanent magnet alternator allowing it to generate power at low RPMs. The Five Gallon Bucket Hydroelectric Generator was the subject of a work group at the 2008 International Development Design Summit (IDDS) at the Massachusetts Institute of Technology. During the Summer of 2013, an energy project in Abra Malaga, Peru was completed using the bucket generator.

A website has been put together as a forum for ideas and further iterations of the generator that includes a build manual. Common centrifugal water pumps, can be operated in reverse to act as turbines. While these machines rarely have optimum hydraulic characteristics when operated as turbines, their availability and low cost makes them attractive.

### Transmission Distance

If the power will be used more than 30 meters (100 feet) from the generator, then the transmission distance may be an important consideration. Many small systems use automotive alternators producing 12 VDC, and possibly charging a battery. For example, a 12 V system that produces 1 kW of power has a flow of 80 A and the wire size is 4-gauge. The purchase cost of two strands of wire 1000 feet long for this voltage and amperage was typically $2400(US) in 2015. To avoid such a large wire cost a higher voltage and lower current is required. If a 240 VAC alternator is used instead the flow is only 4 A over 1000' of 18 gauge wire costing $180(US). The cost of wire resulted in North America using 120/240 VAC after DC voltage lost the war of the currents in the late 1800s. Another approach to reduce wire costs is to have a 12 VDC alternator with a short high current connection to an inverter outputting 120 VAC or 240 VAC at a much lower current on a long length of thinner wire.

## Marine Current Power

Marine currents in rivers, straights and the oceans constitute a renewable energy source with good potential. At Uppsala University, research is primarily focused on the generators of marine current energy converters. The goal is to have a power plants that generate electricity with high efficiency at slow water movement.

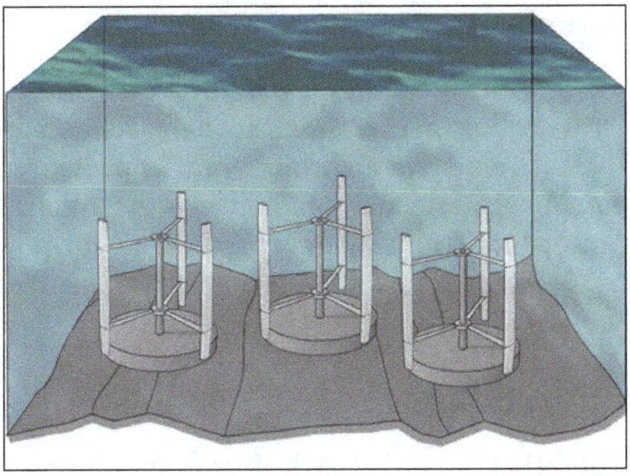

In March 2007, the first prototype generator adapted to these low velocities was completed at the Ångström Laboratory. 2013 the first experimental facility was launched at Soderfors in the Dal river. The next step in the research is to connect the power plants to the grid and to develop and scale up the technology so that it is suitable for tidal applications.

## Tidal Stream Generator

Tidal stream generators make use of the kinetic energy of moving water to turn a turbine—similar to the way a wind turbine uses wind to create electricity. However, the power available for tidal power generation in a given area can be greater than a wind turbine due to the higher density of water. These types of tidal generators tend to be the cheapest (though still quite expensive) and most environmentally friendly of any type of tidal power generation.

These types of generators have a very low visual impact and are mostly or totally submerged. Furthermore, they are less intrusive to marine life because they produce less noise pollution. Tides are more predictable than wind (which varies a great deal). The density of water is also much higher than air, which means that tidal turbines can be much smaller than wind turbines for the same output. This is further enhanced by the use of the Venturi effect, which is a way of getting water to move faster through these turbines.

The initial investment costs for all types of tidal power technology are extremely high and it takes quite a while to make back that investment. In addition, sea water is quite corrosive which leads to high maintenance costs.

Since the tides flow-in and recede once per day, tidal stream generators provide intermittent power generation.

## Varying Types of Generators

Different designs of tidal power generation may prove to offer different advantages, extensive studies are still needed on this relatively new technology. Studies are needed for maintenance costs vs. income and environmental impactsover an extended period of time. Power output can vary considerably between these classes of turbines. However, since the tidal power industry is so new predictions for power output vary considerably. Currently, research data on stream generators is scarce due to the emerging nature of the industry.

## Axial (Horizontal) Turbines

Axial turbines have a rotor that is parallel to the incoming water stream. They use rotors similar to those seen on wind turbines but have modifications due to the differing fluid properties of water from air.

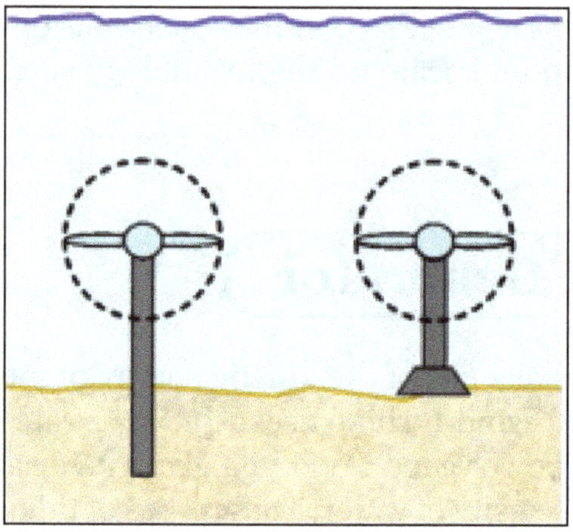

Axial turbines underwater.

These devices use the lift of water to generate power. This requires specially shaped airfoil surfaces designed to create a pressure difference. This leads to a net force in the direction perpendicular to the water flow and thus turns the device. Rotors of this type must be carefully oriented (the orientation is referred to as the rotor pitch), to maintain their ability to harness the power of the tide as it changes.

## Cross-flow Turbines

Cross flow turbines have a rotor that is perpendicular to the water flow but parallel to the water surface. Unlike axial turbines, these devices typically use the drag of water to generate power. In drag-based turbines, the force of the water pushes against a surface, like wind on an open sail. This works because the drag of the open face of the turbine blade is greater than the drag on its closed face. Drag based devices are inherently less efficient than their lift-based counterparts because they operate with respect to the relative flow speed of the water. This means that as a turbine rotates faster the relative flow speed of the water will decrease and thus transfer less of its energy.

A cross flow water turbine.

## Vertical Axis Turbines

With vertical axis turbines the rotational axis of the rotor is vertical to the water surface and also perpendicular to the incoming water flow. Both lift and drag type blades can be used in these turbines.

## Oscillating Generators

Oscillating devices do not have a rotational component, instead these generators make use of sections which are pushed sideways by the flowing current to create a hydraulic pump. This pump transfers its energy to a motor, which then turns a generator and creates electricity.

## References


- Hydro-power, topics: studentenergy.org, Retrieved 18 April, 2020
- Watermill: top-alternative-energy-sources.com, Retrieved 12 February, 2020
- Rehman, Shafiqur; Al-Hadhrami, Luai; Alam, Md (30 April 2015). "Pumped hydro energy storage system: A technological review". Renewable and Sustainable Energy Reviews. 44: 586–598. doi:10.1016/j.rser.2014.12.040 – via ResearchGate
- Marine-currents, electricity-research-areas: teknik.uu.se, Retrieved 22 July, 2020
- Bent Sørensen (2004). Renewable Energy: Its Physics, Engineering, Use, Environmental Impacts, Economy, and Planning Aspects. Academic Press. pp. 556–. ISBN 978-0-12-656153-1
- Tidal-stream-generator, encyclopedia: energyeducation.ca, Retrieved 15 June, 2020

# Bioenergy 5

- Biofuels
- Biomass
- Energy Forestry
- Biodiesel

Bioenergy is a form of renewable energy that generates electricity from organic matter. Biofuels, biodiesel, biogasoline, biorefinery, etc. come under the subject of bioenergy. All the aspects related to bioenergy have been carefully analyzed in this chapter.

Bioenergy is renewable energy created from natural, biological sources. Many natural sources, such as plants, animals, and their byproducts, can be valuable resources. Modern technology even makes landfills or waste zones potential bioenergy resources. It can be used to be a sustainable power source, providing heat, gas, and fuel.

Because the energy contained in sources like plants is obtained from the sun through photosynthesis, it can be replenished and is considered an inexhaustible source.

Using bioenergy has the potential to decrease our carbon footprint and improve the environment. While bioenergy uses the same amount of carbon dioxide as traditional fossil fuels, the impact can be minimized as long as the plants used are replaced. Fast-growing trees and grass help this process and are known as bioenergy feedstocks.

## Sources of Bioenergy

Most bioenergy comes from forests, agricultural farms, and waste. The feedstocks are grown by farms specifically for their use as an energy source. Common crops include starch or sugar-based plants, like sugarcane or corn.

## How it is Created?

To turn raw sources into energy, there are three processes: chemical, thermal, and biochemical. Chemical processing uses chemical agents to break down the natural source and convert it into liquid fuel. Corn ethanol, a fuel created from corn, is an example of chemical processing results. The thermal conversion uses heat to change the source into energy through combustion or gasification. A biochemical conversion uses bacteria or other organisms to convert the source, such as through composting or fermentation.

## Uses of Bioenergy

Bioenergy exists at several different levels. Individuals can create bioenergy, such as by creating a compost heap out of kitchen scraps and keeping worms to produce rich fertilizer. At the other extreme are large energy corporations looking for more sustainable energy sources than oil or coal. These organizations use huge farms and facilities to provides hundreds or thousands of customers with energy.

## Importance of Bioenergy

Having the ability to produce energy through plants or other resources can lessen U.S. reliance on foreign nations for sources of energy. Bioenergy also is viewed as essential for the environment. Continued use of fossil fuels can cause significant environmental issues by producing greenhouse gases that contribute to global warming or by emitting harmful pollutants such as sulfur dioxide which can harm the population's health.

As technology progresses, bioenergy has the potential to dramatically reduce greenhouse emissions, the release of harmful gases associated with global warming and climate change. The use of forests and farms in bioenergy can help combat the harmful release of carbon dioxide and help achieve a balance.

At this time, bioenergy is not ready to replace fossil fuels. The process is too costly and uses too many resources to be practical in most areas. The large plots of land and significant amounts of water needed to be successful can be difficult for many states or countries. Additionally, agricultural resources such as land and water dedicated to producing crops related to bioenergy can limit the resources used for producing food. Still, as science continues to study this area, bioenergy could increasingly become a larger source of energy that can help improve the environment.

## Biofuels

A bus fueled by biogas.

A biofuel is a fuel that is produced through contemporary processes from biomass, rather than a fuel produced by the very slow geological processes involved in the formation of fossil fuels, such as oil. Since biomass technically can be used as a fuel directly (e.g. wood logs), some people use the terms biomass and biofuel interchangeably. More often than not however, the word biomass simply denotes the biological raw material the fuel is made of, or some form of thermally/chemically altered solid end product, like torrefied pellets or briquettes. The word biofuel is usually reserved for liquid or gaseous fuels, used for transportation. The EIA (U.S. Energy Information Administration) follow this naming practice. If the biomass used in the production of biofuel can regrow quickly, the fuel is generally considered to be a form of renewable energy.

Biofuel logo.

Biofuels can be produced from plants (i.e. energy crops), or from agricultural, commercial, domestic, industrial wastes (if the waste has a biological origin). Renewable biofuels generally involve contemporary carbon fixation, such as those that occur in plants or microalgae through the process of photosynthesis.

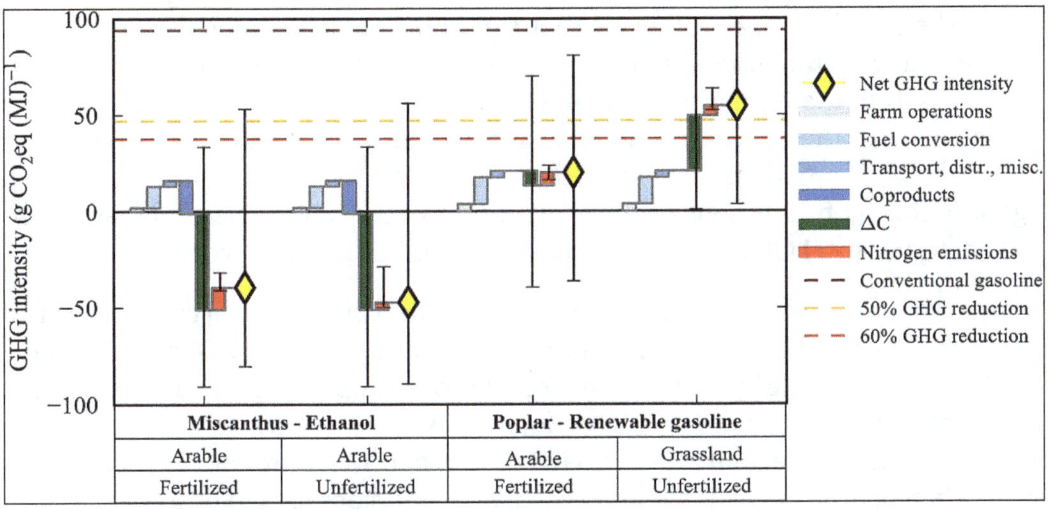


GHG/$CO_2$/carbon negativity for Miscanthus x giganteus production pathways.

Some argue that biofuel can be carbon-neutral because all biomass crops sequester carbon to a certain extent – basically all crops move $CO_2$ from above-ground circulation to below-ground storage in the roots and the surrounding soil. For instance, McCalmont et al. found below-ground carbon accumulation ranging from 0.42 to 3.8 tonnes per hectare per year for soils below Miscanthus x giganteus energy crops, with a mean accumulation rate of 1.84 tonne (0.74 tonnes per acre per year), or 20% of total harvested carbon per year.

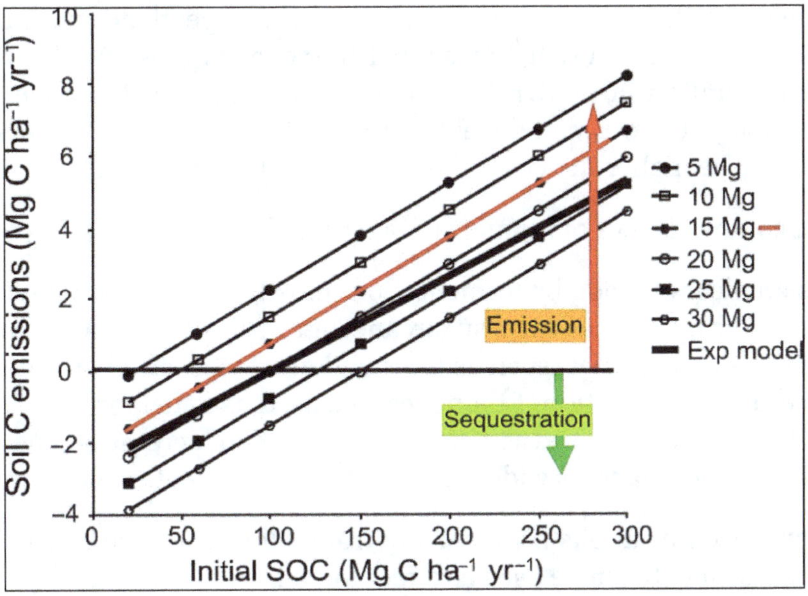

Relationship between above-ground yield (diagonal lines), soil organic carbon (X axis), and soil's potential for successful/unsuccessful carbon sequestration (Y axis). Basically, the higher the yield, the more land is usable as a GHG mitigation tool (including relatively carbon rich land.)

However, the simple proposal that biofuel is carbon-neutral almost by definition has been superseded by the more nuanced proposal that for a particular biofuel project to be carbon neutral, the total carbon sequestered by the energy crop's root system must compensate for all the above-ground emissions (related to this particular biofuel project). This includes any emissions caused by direct or indirect land use change. Many first generation biofuel projects are not carbon neutral given these demands. Some have even higher total GHG emissions than some fossil based alternatives.

Some are carbon neutral or even negative, though, especially perennial crops. The amount of carbon sequestrated and the amount of GHG (greenhouse gases) emitted will determine if the total GHG life cycle cost of a biofuel project is positive, neutral or negative. A carbon negative life cycle is possible if the total below-ground carbon accumulation more than compensates for the total life-cycle GHG emissions above ground. In other words, to achieve carbon neutrality yields should be high and emissions should be low.

High-yielding energy crops are thus prime candidates for carbon neutrality. The graphic on the right displays two $CO_2$ negative Miscanthus x giganteus production pathways, represented in gram $CO_2$-equivalents per megajoule. The yellow diamonds represent mean values. Further, successful sequestration is dependent on planting sites, as the best soils for sequestration are those that are currently low in carbon. The varied results displayed in the graph highlights this fact. For the UK, successful sequestration is expected for arable land over most of England and Wales, with unsuccessful sequestration expected in parts of Scotland, due to already carbon rich soils (existing woodland) plus lower yields. Soils already rich in carbon includes peatland and mature forest. Grassland can also be carbon rich, however Milner et al. argues that the most successful carbon sequestration in the UK takes place below improved grasslands. The bottom graphic displays the estimated yield necessary to compensate for related lifecycle GHG-emissions. The higher the yield, the more likely $CO_2$ negativity becomes.

The two most common types of biofuel are bioethanol and biodiesel.

Bioethanol is an alcohol made by fermentation, mostly from carbohydrates produced in sugar or starch crops such as corn, sugarcane, or sweet sorghum. Cellulosic biomass, derived from non-food sources, such as trees and grasses, is also being developed as a feedstock for ethanol production. Ethanol can be used as a fuel for vehicles in its pure form (E100), but it is usually used as a gasoline additive to increase octane and improve vehicle emissions. Bioethanol is widely used in the United States and in Brazil.

Biodiesel is produced from oils or fats using transesterification and is the most common biofuel in Europe. It can be used as a fuel for vehicles in its pure form (B100), but it is usually used as a diesel additive to reduce levels of particulates, carbon monoxide, and hydrocarbons from diesel-powered vehicles.

In 2018, worldwide biofuel production reached 152 billion liters (40 billion gallons US), up 7% from 2017, and biofuels provided 3% of the world's fuels for road transport. The International Energy Agency want biofuels to meet more than a quarter of world demand for transportation fuels by 2050, in order to reduce dependency on petroleum. However, the production and consumption of biofuels are not on track to meet the IEA's sustainable development scenario. From 2020 to 2030 global biofuel output has to increase by 10% each year to reach IEA's goal. Only 3% growth annually is expected.

Here are some various social, economic, environmental and technical issues relating to biofuels production and use, which have been debated in the popular media and scientific journals.

## Generations

### First-generation Biofuels

"First-generation" or conventional biofuels are biofuels made from food crops grown on arable land. With this biofuel production generation, food crops are thus explicitly grown for fuel production, and not anything else. The sugar, starch, or vegetable oil obtained from the crops is converted into biodiesel or ethanol, using transesterification, or yeast fermentation.

### Second-generation Biofuels

Second generation biofuels are fuels manufactured from various types of biomass. Biomass is a wide-ranging term meaning any source of organic carbon that is renewed rapidly as part of the carbon cycle. Biomass is derived from plant materials, but can also include animal materials.

Whereas first generation biofuels are made from the sugars and vegetable oils found in arable crops, second generation biofuels are made from lignocellulosic biomass or woody crops, agricultural residues or waste plant material (from food crops that have already fulfilled their food purpose). The feedstock used to generate second-generation biofuels thus either grows on arable lands, but are just byproducts of the actual harvest (main crop) or they are grown on lands which cannot be used to effectively grow food crops and in some cases neither extra water or fertilizer is applied to them. Non-human food second generation feedstock sources include grasses, jatropha and other seed crops, waste vegetable oil, municipal solid waste and so forth.

This has both advantages and disadvantages. The advantage is that, unlike with regular food crops, no arable land is used solely for the production of fuel. The disadvantage is that unlike with regular food crops, it may be rather difficult to extract the fuel. For instance, a series of physical and chemical treatments might be required to convert lignocellulosic biomass to liquid fuels suitable for transportation.

## Third-generation Biofuels

From 1978 to 1996, the US NREL experimented with using algae as a biofuels source in the Aquatic Species Program. A self-published article by Michael Briggs, at the UNH Biofuels Group, offers estimates for the realistic replacement of all vehicular fuel with biofuels by using algae that have a natural oil content greater than 50%, which Briggs suggests can be grown on algae ponds at wastewater treatment plants. This oil-rich algae can then be extracted from the system and processed into biofuels, with the dried remainder further reprocessed to create ethanol. The production of algae to harvest oil for biofuels has not yet been undertaken on a commercial scale, but feasibility studies have been conducted to arrive at the above yield estimate. In addition to its projected high yield, algaculture – unlike crop-based biofuels – does not entail a decrease in food production, since it requires neither farmland nor fresh water. Many companies are pursuing algae bioreactors for various purposes, including scaling up biofuels production to commercial levels. Prof. Rodrigo E. Teixeira from the University of Alabama in Huntsville demonstrated the extraction of biofuels lipids from wet algae using a simple and economical reaction in ionic liquids.

## Fourth-generation Biofuels

Similarly to third-generation biofuels, fourth-generation biofuels are made using non-arable land. However, unlike third-generation biofuels, they do not require the destruction of biomass. This class of biofuels includes electrofuels and photobiological solar fuels. Some of these fuels are carbon-neutral.

## Types

The following fuels can be produced using first, second, third or fourth-generation biofuel production procedures. Most of these can even be produced using two or three of the different biofuel generation procedures.

## Syngas

Syngas, a mixture of carbon monoxide, hydrogen and other hydrocarbons, is produced by partial combustion of biomass, that is, combustion with an amount of oxygen that is not sufficient to convert the biomass completely to carbon dioxide and water. Before partial combustion, the biomass is dried, and sometimes pyrolysed. The resulting gas mixture, syngas, is more efficient than direct combustion of the original biofuel; more of the energy contained in the fuel is extracted.

Syngas may be burned directly in internal combustion engines, turbines or high-temperature fuel cells. The wood gas generator, a wood-fueled gasification reactor, can be connected to an internal combustion engine.

Syngas can be used to produce methanol, DME and hydrogen, or converted via the Fischer-Tropsch process to produce a diesel substitute, or a mixture of alcohols that can be blended into gasoline. Gasification normally relies on temperatures greater than 700 °C.

Lower-temperature gasification is desirable when co-producing biochar, but results in syngas polluted with tar.

## Ethanol

Biologically produced alcohols, most commonly ethanol, and less commonly propanol and butanol, are produced by the action of microorganisms and enzymes through the fermentation of sugars or starches (easiest), or cellulose (which is more difficult). Biobutanol (also called biogasoline) is often claimed to provide a direct replacement for gasoline, because it can be used directly in a gasoline engine.

Neat ethanol on the left (A), gasoline on the right (G) at a filling station.

Ethanol fuel is the most common biofuel worldwide, particularly in Brazil. Alcohol fuels are produced by fermentation of sugars derived from wheat, corn, sugar beets, sugar cane, molasses and any sugar or starch from which alcoholic beverages such as whiskey, can be made (such as potato and fruit waste, etc.). The ethanol production methods used are enzyme digestion (to release sugars from stored starches), fermentation of the sugars, distillation and drying. The distillation process requires significant energy input for heat (sometimes unsustainable natural gas fossil fuel, but cellulosic biomass such as bagasse, the waste left after sugar cane is pressed to extract its juice, is the most common fuel in Brazil, while pellets, wood chips and also waste heat are more common in Europe) Waste steam fuels ethanol factory – where waste heat from the factories also is used in the district heating grid.

Ethanol can be used in petrol engines as a replacement for gasoline; it can be mixed with gasoline to any percentage. Most existing car petrol engines can run on blends of up to 15% bioethanol with petroleum/gasoline. Ethanol has a smaller energy density than that of gasoline; this means it takes more fuel (volume and mass) to produce the same amount of work. An advantage of ethanol ($CH_3CH_2OH$) is that it has a higher octane rating than ethanol-free gasoline available at roadside gas stations, which allows an increase of an engine's compression ratio for increased thermal efficiency. In high-altitude (thin air) locations, some states mandate a mix of gasoline and ethanol as a winter oxidizer to reduce atmospheric pollution emissions.

Ethanol is also used to fuel bioethanol fireplaces. As they do not require a chimney and are "flueless", bioethanol fires are extremely useful for newly built homes and apartments without a flue. The downsides to these fireplaces is that their heat output is slightly less than electric heat or gas fires, and precautions must be taken to avoid carbon monoxide poisoning.

Corn-to-ethanol and other food stocks has led to the development of cellulosic ethanol. According to a joint research agenda conducted through the US Department of Energy, the fossil energy ratios (FER) for cellulosic ethanol, corn ethanol, and gasoline are 10.3, 1.36, and 0.81, respectively.

Ethanol has roughly one-third lower energy content per unit of volume compared to gasoline. This is partly counteracted by the better efficiency when using ethanol (in a long-term test of more than 2.1 million km, the BEST project found FFV vehicles to be 1–26% more energy efficient than petrol cars, but the volumetric consumption increases by approximately 30%, so more fuel stops are required).

With current subsidies, ethanol fuel is slightly cheaper per distance traveled in the United States.

## Other Bioalcohols

Methanol is currently produced from natural gas, a non-renewable fossil fuel. In the future it is hoped to be produced from biomass as biomethanol. This is technically feasible, but the production is currently being postponed for concerns that the economic viability is still pending. The methanol economy is an alternative to the hydrogen economy, compared to today's hydrogen production from natural gas.

Butanol ($C_4H_9OH$) is formed by ABE fermentation (acetone, butanol, ethanol) and experimental modifications of the process show potentially high net energy gains with butanol as the only liquid product. Butanol will produce more energy and allegedly can be burned "straight" in existing gasoline engines (without modification to the engine or car), and is less corrosive and less water-soluble than ethanol, and could be distributed via existing infrastructures. DuPont and BP are working together to help develop butanol. Escherichia coli strains have also been successfully engineered to produce butanol by modifying their amino acid metabolism. One drawback to butanol production in E. coli remains the high cost of nutrient rich media, however, recent work has demonstrated E. coli can produce butanol with minimal nutritional supplementation.

## Green Diesel

Green diesel is produced through hydrocracking biological oil feedstocks, such as vegetable oils and animal fats. Hydrocracking is a refinery method that uses elevated temperatures and pressure in the presence of a catalyst to break down larger molecules, such as those found in vegetable oils, into shorter hydrocarbon chains used in diesel engines. It may also be called renewable diesel, hydrotreated vegetable oil (HVO fuel) or hydrogen-derived renewable diesel. Unlike biodiesel, green diesel has exactly the same chemical properties as petroleum-based diesel. It does not require new engines, pipelines or infrastructure to distribute and use, but has not been produced at a cost that is competitive with petroleum. Gasoline versions are also being developed. Green diesel is being developed in Louisiana and Singapore by ConocoPhillips, Neste Oil, Valero, Dynamic Fuels, and Honeywell UOP as well as Preem in Gothenburg, Sweden, creating what is known as Evolution Diesel.

## Straight Vegetable Oil

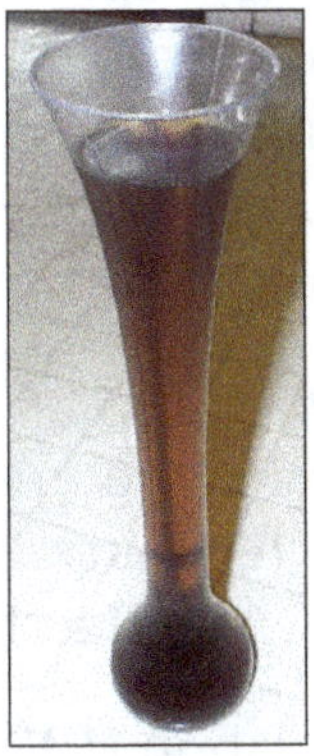

Filtered waste vegetable oil.

Straight unmodified edible vegetable oil is generally not used as fuel, but lower-quality oil has been used for this purpose. Used vegetable oil is increasingly being processed into biodiesel, or (more rarely) cleaned of water and particulates and then used as a fuel.

As with 100% biodiesel (B100), to ensure the fuel injectors atomize the vegetable oil in the correct pattern for efficient combustion, vegetable oil fuel must be heated to reduce its viscosity to that of diesel, either by electric coils or heat exchangers. This is easier in warm or temperate climates. MAN B&W Diesel, Wärtsilä, and Deutz AG, as well as a number of smaller companies, such as Elsbett, offer engines that are compatible with straight vegetable oil, without the need for after-market modifications.

Vegetable oil can also be used in many older diesel engines that do not use common rail or unit injection electronic diesel injection systems. Due to the design of the combustion chambers in indirect injection engines, these are the best engines for use with vegetable oil. This system allows the relatively larger oil molecules more time to burn. Some older engines, especially Mercedes, are driven experimentally by enthusiasts without any conversion, a handful of drivers have experienced limited success with earlier pre-"Pumpe Duse" VW TDI engines and other similar engines with direct injection. Several companies, such as Elsbett or Wolf, have developed professional conversion kits and successfully installed hundreds of them over the last decades.

Oils and fats can be hydrogenated to give a diesel substitute. The resulting product is a straight-chain hydrocarbon with a high cetane number, low in aromatics and sulfur and does not contain oxygen. Hydrogenated oils can be blended with diesel in all proportions. They have several advantages over biodiesel, including good performance at low temperatures, no storage stability problems and no susceptibility to microbial attack.

## Bioethers

Bioethers (also referred to as fuel ethers or oxygenated fuels) are cost-effective compounds that act as octane rating enhancers."Bioethers are produced by the reaction of reactive iso-olefins, such as iso-butylene, with bioethanol." Bioethers are created from wheat or sugar beets. They also enhance engine performance, while significantly reducing engine wear and toxic exhaust emissions. Although bioethers are likely to replace petroethers in the UK, it is highly unlikely they will become a fuel in and of itself due to the low energy density. By greatly reducing the amount of ground-level ozone emissions, they contribute to air quality.

When it comes to transportation fuel there are six ether additives: dimethyl ether (DME), diethyl ether (DEE), methyl tert-butyl ether (MTBE), ethyl tert-butyl ether (ETBE), tert-amyl methyl ether (TAME), and tert-amyl ethyl ether (TAEE).

The European Fuel Oxygenates Association (EFOA) identifies methyl tert-butyl ether (MTBE) and ethyl tert-butyl ether (ETBE) as the most commonly used ethers in fuel to replace lead. Ethers were introduced in Europe in the 1970s to replace the highly toxic compound. Although Europeans still use bioether additives, the US no longer has an oxygenate requirement therefore bioethers are no longer used as the main fuel additive.

## Compatibility with Existing Infrastructure

So-called "drop-in" biofuels can be defined as "liquid bio-hydrocarbons that are functionally equivalent to petroleum fuels and are fully compatible with existing petroleum infrastructure". Drop-in biofuels require no (engine) modification of the vehicle.

Some examples of drop-in biofuels include biobutanol, biodiesel, synthetic paraffinic kerosine, and other synthetic fuels.

According to a July 2014 report published by the IEA Bioenergy Task 39, entitled "The Potential and Challenges of Drop-in Biofuels", there are several ways to produce drop-in biofuels that are functionally equivalent to petroleum-derived transportation fuel blendstocks. These are discussed within three major sections of the full report and include:

- Oleochemical processes, such as the hydroprocessing of lipid feedstocks obtained from oilseed crops, algae or tallow;
- Thermochemical processes, such as the thermochemical conversion of biomass to fluid intermediates (gas or oil) followed by catalytic upgrading and hydroprocessing to hydrocarbon fuels;
- Biochemical processes, such as the biological conversion of biomass (sugars, starches or lignocellulose-derived feedstocks) to longer chain alcohols and hydrocarbons.

A fourth category is also briefly described that includes "hybrid" thermochemical/biochemical technologies such as fermentation of synthesis gas and catalytic reforming of sugars/carbohydrates.

The report concludes by stating:

> Tremendous entrepreneurial activity to develop and commercialize drop-in biofuels from aquatic and terrestrial feedstocks has taken place over the past several years. However, despite these efforts, drop-in biofuels represent only a small percentage (around 2%) of global biofuel markets. Due to the increased processing and resource requirements (e.g., hydrogen and catalysts) needed to make drop-in biofuels as compared to conventional biofuels, large scale production of cost-competitive drop-in biofuels is not expected to occur in the near to midterm. Rather, dedicated policies to promote development and commercialization of these fuels will be needed before they become significant contributors to global biofuels production. Currently, no policies (e.g., tax breaks, subsidies etc.) differentiate new, more fungible and infrastructure ready drop-in type biofuels from less infrastructure compatible oxygenated biofuels. Thus, while tremendous technical progress has been made in developing and improving the various routes to drop-in fuels, supportive policies directed specifically towards the further development of drop-in biofuels are likely to be needed to ensure their future commercial success.

## Air Pollution

Biofuels are similar to fossil fuels in that biofuels contribute to air pollution. Burning produces carbon dioxide, airborne carbon particulates, carbon monoxide and nitrous oxides. The WHO estimates 3.7 million premature deaths worldwide in 2012 due to air pollution. Brazil burns significant amounts of ethanol biofuel. Gas chromatograph studies were performed of ambient air in São Paulo, Brazil, and compared to Osaka, Japan, which does not burn ethanol fuel. Atmospheric Formaldehyde was 160% higher in Brazil, and Acetaldehyde was 260% higher.

The Environmental Protection Agency acknowledged in April 2007 that the increased use of bioethanol will lead to worse air quality. The total emissions of air pollutants such as nitrogen oxides will rise due the growing use of bioethanol. There is an increase in carbon dioxide from the burning of fossil fuels to produce the biofuels as well as nitrous oxide from the soil, which has most likely been treated with nitrogen fertilizer. Nitrous oxide is known to have a greater impact on the atmosphere in relation to global warming, as it is also an ozone destroyer.

## Debates Regarding the Production and use of Biofuel

There are various social, economic, environmental and technical issues with biofuel production and use, which have been discussed in the popular media and scientific journals. These include: the effect of moderating oil prices, the "food vs fuel" debate, food prices, poverty reduction potential, energy ratio, energy requirements, carbon emissions levels, sustainable biofuel production, deforestation and soil erosion, loss of biodiversity, impact on water resources, the possible modifications necessary to run the engine on biofuel, as well as energy balance and efficiency. The International Resource Panel, which provides independent scientific assessments and expert advice on a variety of resource-related themes, assessed the issues relating to biofuel use in its first report Towards sustainable production and use of resources: Assessing Biofuels. "Assessing Biofuels" outlined the wider and interrelated factors that need to be considered when deciding on the relative merits of pursuing one biofuel over another. It concluded that not all biofuels perform equally in terms of their impact on climate, energy security and ecosystems, and suggested that environmental and social impacts need to be assessed throughout the entire life-cycle.

Another issue with biofuel use and production is the US has changed mandates many times because the production has been taking longer than expected. The Renewable Fuel Standard (RFS) set by congress for 2010 was pushed back to at best 2012 to produce 100 million gallons of pure ethanol (not blended with a fossil fuel).

## Banning of First-generation Biofuels

In the EU, the revised renewable energy directive calls for a complete ban on first-generation biofuels by 2020, as was decided by the EU Parliament in 2016. Particularly fuels made from such oils such as palm oil and soy oil are being targeted.

## Sustainable Biofuels

Many of the biofuels that were being supplied in 2008 (using the first-generation biofuel production procedure) have been criticised for their adverse impacts on the natural environment, food security, and land use. In 2008, the Nobel-prize winning chemist Paul J. Crutzen published findings that the release of nitrous oxide ($N_2O$) emissions in the production of biofuels means that overall they contribute more to global warming than the fossil fuels they replace. In 2008, the challenge was to support biofuel development, including the development of new cellulosic technologies, with responsible policies and economic instruments to help ensure that biofuel commercialization is sustainable. Responsible commercialization of biofuels represented an opportunity to enhance sustainable economic prospects in Africa, Latin America and Asia. Biofuels in the form of liquid fuels derived from plant materials have entered the market, driven by the perception that they reduce climate gas emissions, and also by factors such as oil price spikes and the need for increased energy security.

According to the Rocky Mountain Institute, sound biofuel production practices would not hamper food and fibre production, nor cause water or environmental problems, and would enhance soil fertility. The selection of land on which to grow the feedstocks is a critical component of the ability of biofuels to deliver sustainable solutions. A key consideration is the minimisation of biofuel competition for prime cropland.

## Greenhouse Gas Emissions

Some scientists have expressed concerns about land-use change in response to greater demand for crops to use for biofuel and the subsequent carbon emissions. The payback period, that is, the time it will take biofuels to pay back the carbon debt they acquire due to land-use change, has been estimated to be between 100 and 1000 years, depending on the specific instance and location of land-use change. However, no-till practices combined with cover-crop practices can reduce the payback period to three years for grassland conversion and 14 years for forest conversion.

A study conducted in the Tocantis State, in northern Brazil, found that many families were cutting down forests in order to produce two conglomerates of oilseed plants, the J. curcas (JC group) and the R. communis (RC group). This region is composed of 15% Amazonian rainforest with high biodiversity, and 80% Cerrado forest with lower biodiversity. During the study, the farmers that planted the JC group released over 2193 Mg $CO_2$, while losing 53-105 Mg $CO_2$ sequestration from deforestation; and the RC group farmers released 562 Mg $CO_2$, while losing 48-90 Mg $CO_2$ to be sequestered from forest depletion. The production of these types of biofuels not only led into an increased emission of carbon dioxide, but also to lower efficiency of forests to absorb the gases that these farms were emitting. This has to do with the amount of fossil fuel the production of fuel crops involves. In addition, the intensive use of monocropping agriculture requires large amounts of water irrigation, as well as of fertilizers, herbicides and pesticides. This does not only lead to poisonous chemicals to disperse on water runoff, but also to the emission of nitrous oxide ($NO_2$) as a fertilizer byproduct, which is three hundred times more efficient in producing a greenhouse effect than carbon dioxide ($CO_2$).

Converting rainforests, peatlands, savannas, or grasslands to produce food crop–based biofuels in Brazil, Southeast Asia, and the United States creates a "biofuel carbon debt" by releasing 17 to 420 times more $CO_2$ than the annual greenhouse gas (GHG) reductions that these biofuels would provide by displacing fossil fuels. Biofuels made from waste biomass or from biomass grown on abandoned agricultural lands incur little to no carbon debt.

In addition to crop growth requiring water, biofuel facilities require significant process water.

## Disadvantages of Biofuels

## Regional Suitability

This one relates to the little multicolored maps on the backs of seed packets. The ragged stripes stretching from east to west are growing zones: regions where water supply, temperature and sunlight make hospitable climates for certain types of plants. If you live in Zone 5, for example, you will likely have trouble growing a plant that requires the long growing season and high heat of Zone 10.

Biofuel crops are no different from petunias or peppers in this regard. Certain crops will grow better in certain regions and may not grow at all in others. And while the range of oil-producing crops considered viable for biofuel production is wide enough to fit most growing zones, the most productive crops simply won't grow everywhere. Consumers living in a low-producing region would need to have biofuel trucked or piped to them, increasing both cost and the amount of emissions produced in production and transport.

Researchers are working to increase biofuel yields from weather-tolerant crops. But in much the same way that oranges will never be a cash crop in Alaska, there will always be some regions that simply can't support large-scale production of biofuel-rich crops.

## Water Use

Ask any grade-school student what a plant needs to grow, and he or she will likely mention two things: sunlight and water. While the first is a bit beyond the control of biofuel producers, the second is at the core of a potentially serious drawback of plant-based fuels: The water demands of some biofuel-producing crops could put unsustainable pressure on local water resources if not managed wisely.

A 2009 study suggests that, in the rush to produce enough corn-based ethanol to meet federal alternative energy requirements, biofuel demand is already putting stress on fresh water supplies in the Great Plains and central Southwest. Central to the problem is corn's relatively high water requirement. Researchers are investigating ways to genetically engineer less thirsty crops, and carefully planning what biofuel crops to plant in a given region can mitigate this problem. But large-scale biofuel production especially using corn, and in arid parts of the world will have to share finite water resources with drinking and irrigation needs.

The amount of water need to irrigate biofuel crops like corn can put a strain on local water supplies.

## Food Security

Biofuel production using food crops such as corn, soybeans and sorghum has the potential to alter drastically the world's access to affordable food. The simple supply-and-demand economics of biofuels increase demand for corn, for example, and corn becomes more expensive can pose a threat to some regions' food security, or the access to affordable nutritious food for the region's population.

The rise in demand for food-biofuel crops can have a positive effect for crop producers, in the form of higher prices for their produce. But that price quickly trickles down to consumers. A pig farmer, for example, may have to pay a few extra dollars per bushel to buy corn to feed his livestock. That directly translates into more expensive bacon and ham at the grocery store. For the billions of people who live on only a few dollars per day, even a small increase in food prices could put their access to proper nutrition at risk.

One way to counter this lies in simple diplomacy: The globalization of world commerce means that it's now easier than ever to move food supplies from one part of the world to another in response to increased demand. However, ready access to food imports, and the ease of exporting, hinge on a wide range of political and social factors. Relying on produce from halfway around the globe to feed a hungry nation is a risky price to pay for widespread biofuel integration into the world's energy supplies.

## Deforestation

It seemed like a win-win idea: European demand for biofuel was set to spike, driven in part by regulations aimed at reducing greenhouse gas emissions. Industry researchers had found an answer in palm oil, a relatively easy-to-produce biofuel source. Plantation owners prepared their operations to meet the demand and environmental chaos ensued. According to some estimates, expansion by Indonesian palm oil plantations caused the vast majority of that nation's deforestation in the late '80s and '90s. And high-consumption production practices moving palm oil with petroleum-powered trucks and the practice of draining and burning peat bogs to prepare farmland have made the southeast Asian nation one of the world's leading greenhouse gas emitters.

The growth of palm oil plantations in Indonesia has led to massive amounts of deforestation.

The Indonesian palm oil problem is really a combination of biofuel's drawbacks. The regional nature of high-producing plants such as palm oil means that certain parts of the world are agricultural gold mines: Biofuel demand motivates plantations to expand quickly. But if not done with an eye toward conserving resources and maintaining the spirit of reducing emissions through plant-based fuels, this ramping up of production can lead to greater environmental problems than the ones it's meant to solve.

## Fertilizer Use

This is a problem biofuel crops share with food crops, gardens and lawns worldwide. All of these plants grow better when given fertilizer. But those fertilizers can have harmful effects on the surrounding environment, and expanded biofuel production could mean a major pollution threat to sources of fresh water.

Many fertilizers contain nitrogen and phosphorus. While both of these additives promote rapid and hearty growth in many crops, they have a downside. Overuse or inappropriate application can leave excess fertilizer in the soil, which then washes through regional watersheds and into streams, rivers, lakes and underground aquifers. And once the chemicals are in the water supply, bad things can happen.

Phosphorus has been implicated as a trigger of localized algae blooms: The tiny aquatic plants feed off it and rapidly reproduce, often killing other plants and aquatic animals by reducing the amount of oxygen in water or by releasing toxic chemicals. Nitrogen in drinking water can lead to a host of health problems, including methemoglobinemia, a condition that prevents infants from utilizing the oxygen in their blood. Careful fertilizer application can help prevent widespread pollution problems, but expanding biofuel production to meet the world's demand opens the door for more mistakes in this realm.

## Fuel Use

It might seem counterintuitive at first, but some scientists argue that widespread biofuel production is a negative-sum game: Producing enough biodiesel or ethanol to replace one gallon of petroleum fuel, they argue, requires the energy equivalent to several gallons' worth of petroleum fuel.

While corn may be a viable fuel source, the energy used to turn it into ethanol may not be so environmentally friendly.

To put it another way, think about a field of corn being grown for ethanol. It may produce 100 gallons of the fuel out of one season's crop. But if the tractors that tend the field burn 75 gallons of fuel in the season, the truck to transport the corn to a processor burns 20 gallons on the trip, and the processor uses the energy of 40 gallons of fuel to run its distillation equipment, is the ethanol produced really an environmentally friendly, low-emission fuel? Add other resource costs into the equation, such as the gallons of fresh water needed to grow the plants and the amount of fertilizer needed to keep them healthy, and it becomes even harder to equate biofuel with real energy and carbon emission savings.

A 2005 study suggested that, using current farming and production technology, it takes anywhere from 27 to 118 percent more energy to produce a gallon of biodiesel than the energy it contains. While technology may eventually narrow those ratios, the input-output energy ratio of modern biofuel production is a major drawback to its widespread use.

## Technical Challenges

Perhaps the most straightforward of biofuel's drawbacks is the most obvious: It isn't petroleum-based fuel, so it will operate differently in engines designed for petroleum-based fuel.

Biofuels can be tricky, because they operate differently in cars than regular petroleum-based fuels do.

Corn-based ethanol, for example, has a higher density than gasoline; fuel injectors have to be larger in an ethanol-only engine to match the fuel flow of a comparable gasoline engine. And alcohol fuels (including ethanol) can corrode or damage some of the metal and rubber fittings used in gasoline-powered engines. The conversion from one fuel to the other, in some cases, requires a range of new injectors, gaskets and fuel lines. And once the engine's running, the differences in combustion properties between gasoline and ethanol means that the ethanol-converted engine needs to have its ignition timing adjusted to operate properly.

Biodiesel doesn't fare much better. Because of the higher-than-petroleum gel point of many biodiesel-producing oils, a biodiesel engine can be difficult if not impossible to start in cold weather. The problem is even worse for pure vegetable oil, used as fuel in so-called "greasecars." Drivers of vehicles using these fuels often have heating units installed to keep the fuel tank and lines free from gelled fuel, or install dual-fuel systems that flush the engine with petroleum diesel on start-up and shut-down. A number of manufacturers sell components for biodiesel and greasecar conversions, and intrepid tinkerers often find ways to overcome the gelling problem. But the conversions add time and money to the biofuel equation, something that can be off-putting for potential biofuel users.

## Biomass

Biomass is plant or animal material used for energy production (electricity or heat), or in various industrial processes as raw substance for a range of products. It can be purposely grown energy crops (e.g. miscanthus, switchgrass), wood or forest residues, waste from food crops (wheat straw, bagasse), horticulture (yard waste), food processing (corn cobs), animal farming (manure, rich in nitrogen and phosphorus), or human waste from sewage plants.

Burning plant-derived biomass releases $CO_2$, but it has still been classified as a renewable energy source in the EU and UN legal frameworks because photosynthesis cycles the $CO_2$ back into new crops. In some cases, this recycling of $CO_2$ from plants to atmosphere and back into plants can even be $CO_2$ negative, as a relatively large portion of the $CO_2$ is moved to the soil during each cycle.

Wood pellets.

Cofiring with biomass has increased in coal power plants, because it makes it possible to release less $CO_2$ without the cost associated with building new infrastructure. Co-firing is not without issues however, often an upgrade of the biomass is beneficiary. Upgrading to higher grade fuels can be achieved by different methods, broadly classified as thermal, chemical, or biochemical.

## Biomass Feedstocks

Historically, humans have harnessed biomass-derived energy since the time when people began burning wood fuel. Even in 2019, biomass is the only source of fuel for domestic use in many developing countries. All biomass is biologically-produced matter based in carbon, hydrogen and oxygen. The estimated biomass production in the world is approximately 100 billion metric tons of carbon per year, about half in the ocean and half on land.

Biomass plant

Wood waste outside biomass power plant.

Wood and residues from wood, for instance spruce, birch, eucalyptus, willow, oil palm, remains the largest biomass energy source today. It is used directly as a fuel or processed into pellet fuel or other forms of fuels. Biomass also includes plant or animal matter that can be converted into fuel, fibers or industrial chemicals. There are numerous types of plants, including corn, switchgrass, miscanthus, hemp, sorghum, sugarcane, and bamboo. The main waste energy feedstocks are wood waste, agricultural waste, municipal solid waste, manufacturing waste, and landfill gas. Sewage sludge is another source of biomass. There is ongoing research involving algae or algae-derived biomass. Other biomass feedstocks are enzymes or bacteria from various sources, grown in cell cultures or hydroponics.

Miscanthus x giganteus energy crop.

Based on the source of biomass, biofuels are classified broadly into two major categories:

First-generation biofuels are derived from food sources, such as sugarcane and corn starch. Sugars present in this biomass are fermented to produce bioethanol, an alcohol fuel which serve as an additive to gasoline, or in a fuel cell to produce electricity.

Second-generation biofuels utilize non-food-based biomass sources such as perennial energy crops (low input crops), and agricultural/municipal waste. There is huge potential for second generation biofuels but the resources are currently under-utilized.

## Biomass Conversion

## Thermal Conversions

Thermal conversion processes use heat as the dominant mechanism to upgrade biomass into a better and more practical fuel. The basic alternatives are torrefaction, pyrolysis, and gasification, these are separated principally by the extent to which the chemical reactions involved are allowed to proceed (mainly controlled by the availability of oxygen and conversion temperature).

There are other less common, more experimental or proprietary thermal processes that may offer benefits, such as hydrothermal upgrading. Some have been developed for use on high moisture content biomass, including aqueous slurries, and allow them to be converted into more convenient forms.

Straw bales.

## Chemical Conversion

A range of chemical processes may be used to convert biomass into other forms, such as to produce a fuel that is more practical to store, transport and use, or to exploit some property of the process itself. Many of these processes are based in large part on similar coal-based processes, such as the Fischer-Tropsch synthesis. Biomass can be converted into multiple commodity chemicals.

## Biochemical Conversion

As biomass is a natural material, many highly efficient biochemical processes have developed in nature to break down the molecules of which biomass is composed, and many of these biochemical conversion processes can be harnessed. In most cases, microorganisms are used to perform the conversion process: anaerobic digestion, fermentation, and composting.

Glycoside hydrolases are the enzymes involved in the degradation of the major fraction of biomass, such as polysaccharides present in starch and lignocellulose. Thermostable variants are gaining increasing roles as catalysts in biorefining applications, since recalcitrant biomass often needs thermal treatment for more efficient degradation.

## Electrochemical Conversion

Biomass can be directly converted to electrical energy via electrochemical (electrocatalytic) oxidation of the material. This can be performed directly in a direct carbon fuel cell, direct liquid fuel cells such as direct ethanol fuel cell, a direct methanol fuel cell, a direct formic acid fuel cell, a L-ascorbic Acid Fuel Cell (vitamin C fuel cell), and a microbial fuel cell. The fuel can also be consumed indirectly via a fuel cell system containing a reformer which converts the biomass into a mixture of CO and $H_2$ before it is consumed in the fuel cell.

## Environmental Impact

On combustion, the carbon from biomass is released into the atmosphere as carbon dioxide ($CO_2$). After a period of time ranging from a few months to decades, the $CO_2$ produced from combustion is absorbed from the atmosphere by plants or trees. However, the carbon storage capacity of forests may be reduced overall if destructive forestry techniques are employed.

All biomass crops sequester carbon. For example, soil organic carbon has been observed to be greater below switchgrass crops than under cultivated cropland, especially at depths below 30 cm (12 in). For Miscanthus x giganteus, McCalmont et al. found accumulation rates ranging from 0.42 to 3.8 tonnes per hectare per year, with a mean accumulation rate of 1.84 tonne (0.74 tonnes per acre per year), or 20% of total harvested carbon per year. The grass sequesters carbon in its continually increasing root biomass, together with carbon input from fallen leaves. Typically, perennial crops sequester more carbon than annual crops because the root buildup is allowed to continue undisturbed over many years. Also, perennial crops avoid the yearly tillage procedures (plowing, digging) associated with growing annual crops. Tilling induces soil aeration, which accelerates the soil carbon decomposition rate, by stimulating soil microbe populations. Also, tilling makes it easier for the oxygen (O) atoms in the atmosphere to attach to carbon (C) atoms in the soil, producing $CO_2$.

The simple proposal that biomass is carbon-neutral put forward in the early 1990s has been superseded by the more nuanced proposal that for a particular bioenergy project to be carbon neutral, the total carbon sequestered by a bioenergy crop's root system must compensate for all the emissions from the related, aboveground bioenergy project. This includes any emissions caused by direct or indirect land use change. Many first generation bioenergy projects are not carbon neutral given these demands. Some have even higher total GHG emissions than some fossil based alternatives. Transport fuels might be worse than solid fuels in this regard.

Some are carbon neutral or even negative, though, especially perennial crops. The amount of carbon sequestrated and the amount of GHG (greenhouse gases) emitted will determine if the total GHG life cycle cost of a bio-energy project is positive, neutral or negative. Specifically, a GHG/carbon negative life cycle is possible if the total below-ground carbon accumulation more than compensates for the above-ground total life-cycle GHG emissions. Whitaker et al. estimates that for Miscanthus x giganteus, carbon neutrality and even negativity is within reach. Basically, the yield and related carbon sequestration is so high that it more than compensates for both farm operations emissions, fuel conversion emissions and transport emissions. The graphic on the right displays two $CO_2$ negative Miscanthus x giganteus production pathways, represented in gram $CO_2$-equivalents per megajoule. The yellow diamonds represent mean values.

One should note that successful sequestration is dependent on planting sites, as the best soils for sequestration are those that are currently low in carbon. The varied results displayed in the graph highlights this fact. Milner et al. argues that for the UK, successful sequestration is expected for arable land over most of England and Wales, with unsuccessful sequestration expected in parts of Scotland, due to already carbon rich soils (existing woodland). Also, for Scotland, the relatively lower yields in this colder climate makes $CO_2$ negativity harder to achieve. Soils already rich in carbon includes peatland and mature forest. Grassland can also be carbon rich, however Milner et al. further argues that the most successful carbon sequestration in the UK takes place below improved grasslands. The bottom graphic displays the estimated yield necessary to achieve $CO_2$ negativity for different levels of existing soil carbon saturation.

Forest-based biomass projects has received criticism for ineffective GHG mitigation from a number of environmental organizations, including Greenpeace and the Natural Resources Defense Council. Environmental groups also argue that it might take decades for the carbon released by burning biomass to be recaptured by new trees. Biomass burning produces air pollution in the form of carbon monoxide, volatile organic compounds, particulates and other pollutants. In 2009 a Swedish study of the giant brown haze that periodically covers large areas in South Asia determined that two thirds of it had been principally produced by residential cooking and agricultural burning, and one third by fossil-fuel burning. The use of wood biomass as an industrial fuel has been shown to produce fewer particulates and other pollutants than the burning seen in wildfires or open field fires.

## Energy Forestry

Energy forestry is a form of forestry in which a fast-growing species of tree or woody shrub is grown specifically to provide biomass or biofuel for heating or power generation.

The two forms of energy forestry are short rotation coppice and short rotation forestry:

- Short rotation coppice may include tree crops of poplar, willow or eucalyptus, grown for two to five years before harvest.
- Short rotation forestry are crops of alder, ash, birch, eucalyptus, poplar, and sycamore, grown for eight to 20 years before harvest.

### Benefits

The main advantage of using "grown fuels", as opposed to fossil fuels such as coal, natural gas and oil, is that while they are growing they absorb the near-equivalent in carbon dioxide (an important greenhouse gas) to that which is later released in their burning.

In comparison, burning fossil fuels increases atmospheric carbon unsustainably, by using carbon that was added to the Earth's carbon sink millions of years ago. This is a prime contributor to climate change.

According to the FAO, compared to other energy crops, wood is among the most efficient sources of bioenergy in terms of quantity of energy released by unit of carbon emitted. Other advantages of generating energy from trees, as opposed to agricultural crops, are that trees do not have to be harvested each year, the harvest can be delayed when market prices are down, and the products can fulfil a variety of end-uses.

Yields of some varieties can be as high as 12 oven dry tonnes every year. However, commercial experience on plantations in Scandinavia have shown lower yield rates.

These crops can also be used in bank stabilisation and phytoremediation. In fact, experiments in Sweden with willow plantations have proved to have many beneficial effects on the soil and water quality when compared to conventional agricultural crops (such as cereal).

### Problems

Although in many areas of the world government funding is still required to support large scale development of energy forestry as an industry, it is seen as a valuable component of the renewable energy network and will be increasingly important in the future.

Growing trees is relatively water intensive.

The system of energy forestry has faced criticism over food vs. fuel, whereby it has become financially profitable to replace food crops with energy crops. It has to be noted, however, that such energy forests do not necessarily compete with food crops for highly productive land as they can be grown on slopes, marginal, or degraded land as well - sometimes even with long-term restoration purposes in mind.

## Biodiesel

Biodiesel refers to a vegetable oil- or animal fat-based diesel fuel consisting of long-chain alkyl (methyl, ethyl, or propyl) esters. Biodiesel is typically made by chemically reacting lipids (e.g., vegetable oil, soybean oil, animal fat (tallow)) with an alcohol producing fatty acid esters.

Biodiesel is a drop-in biofuel and thus meant to be used in standard diesel engines and is thus distinct from the vegetable and waste oils used to fuel converted diesel engines. Biodiesel can be used alone, or blended with petrodiesel in any proportions. Biodiesel blends can also be used as heating oil.

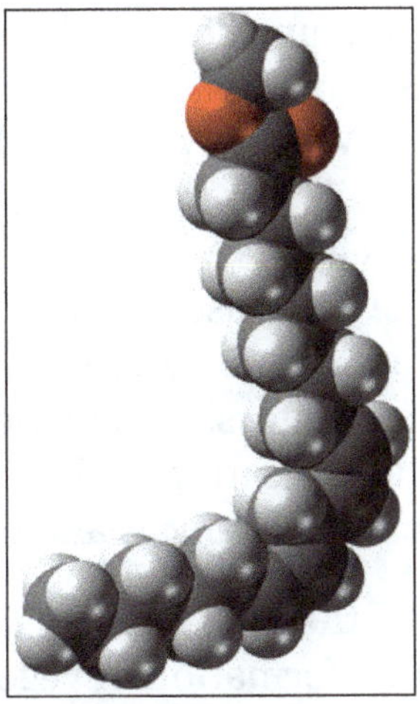

Space-filling model of methyl linoleate, or linoleic acid methyl ester, a common methyl ester produced from soybean or canola oil and methanol.

The National Biodiesel Board (USA) defines "biodiesel" as a mono-alkyl ester.

Transesterification of a vegetable oil was conducted as early as 1853 by Patrick Duffy, four decades before the first diesel engine became functional. Rudolf Diesel's prime model, a single 10 ft (3.05 m) iron cylinder with a flywheel at its base, ran on its own power for the first time in Augsburg, Germany, on 10 August 1893 running on nothing but peanut oil. In remembrance of this event, 10 August has been declared "International Biodiesel Day".

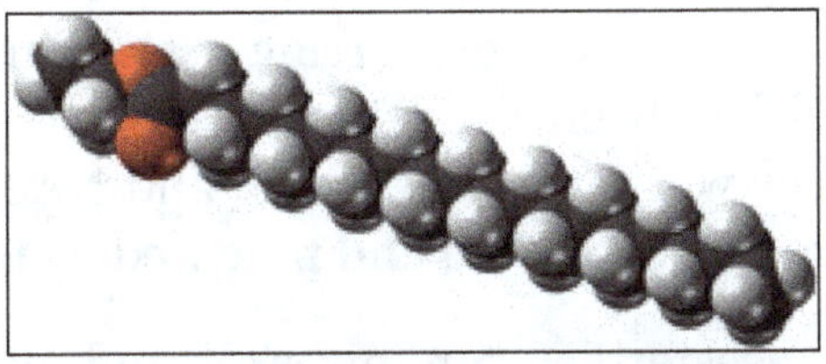

Space-filling model of ethyl stearate, or stearic acid ethyl ester, an ethyl ester produced from soybean or canola oil and ethanol.

It is often reported that Diesel designed his engine to run on peanut oil, but this is not the case. Diesel stated in his published papers, "at the Paris Exhibition in 1900 (Exposition Universelle) there was shown by the Otto Company a small Diesel engine, which, at the request of the French government ran on arachide (earth-nut or pea-nut) oil, and worked so smoothly that only a few people were aware of it. The engine was constructed for using mineral oil, and was then worked on vegetable oil without any alterations being made. The French Government at the time thought of testing the applicability to power production of the Arachide, or earth-nut, which grows in considerable quantities in their African colonies, and can easily be cultivated there." Diesel himself later conducted related tests and appeared supportive of the idea. In a 1912 speech Diesel said, "the use of vegetable oils for engine fuels may seem insignificant today but such oils may become, in the course of time, as important as petroleum and the coal-tar products of the present time."

Despite the widespread use of petroleum-derived diesel fuels, interest in vegetable oils as fuels for internal combustion engines was reported in several countries during the 1920s and 1930s and later during World War II. Belgium, France, Italy, the United Kingdom, Portugal, Germany, Brazil, Argentina, Japan and China were reported to have tested and used vegetable oils as diesel fuels during this time. Some operational problems were reported due to the high viscosity of vegetable oils compared to petroleum diesel fuel, which results in poor atomization of the fuel in the fuel spray and often leads to deposits and coking of the injectors, combustion chamber and valves. Attempts to overcome these problems included heating of the vegetable oil, blending it with petroleum-derived diesel fuel or ethanol, pyrolysis and cracking of the oils.

On 31 August 1937, G. Chavanne of the University of Brussels (Belgium) was granted a patent for a "Procedure for the transformation of vegetable oils for their uses as fuels" (fr. "Procédé de Transformation d'Huiles Végétales en Vue de Leur Utilisation comme Carburants") Belgian Patent 422,877. This patent described the alcoholysis (often referred to as transesterification) of vegetable oils using ethanol (and mentions methanol) in order to separate the fatty acids from the glycerol by replacing the glycerol with short linear alcohols. This appears to be the first account of the production of what is known as "biodiesel" today. This is similar (copy) to the patented methods used in the 18th century to make lamp-oil, and may be inspired by some old historical oil lamps, in some places.

More recently, in 1977, Brazilian scientist Expedito Parente invented and submitted for patent, the first industrial process for the production of biodiesel. This process is classified as biodiesel by international norms, conferring a "standardized identity and quality. No other proposed biofuel has been validated by the motor industry." As of 2010, Parente's company Tecbio is working with Boeing and NASA to certify bioquerosene (bio-kerosene), another product produced and patented by the Brazilian scientist.

Research into the use of transesterified sunflower oil, and refining it to diesel fuel standards, was initiated in South Africa in 1979. By 1983, the process for producing fuel-quality, engine-tested biodiesel was completed and published internationally. An Austrian company, Gaskoks, obtained the technology from the South African Agricultural Engineers; the company erected the first biodiesel pilot plant in November 1987, and the first industrial-scale plant in April 1989 (with a capacity of 30,000 tons of rapeseed per annum).

Throughout the 1990s, plants were opened in many European countries, including the Czech Republic, Germany and Sweden. France launched local production of biodiesel fuel from rapeseed oil, which is mixed into regular diesel fuel at a level of 5%, and into the diesel fuel used by some captive fleets (e.g. public transportation) at a level of 30%. Renault, Peugeot and other manufacturers have certified truck engines for use with up to that level of partial biodiesel; experiments with 50% biodiesel are underway. During the same period, nations in other parts of the world also saw local production of biodiesel starting up: by 1998, the Austrian Biofuels Institute had identified 21 countries with commercial biodiesel projects. 100% biodiesel is now available at many normal service stations across Europe.

## Blends

Blends of biodiesel and conventional hydrocarbon-based diesel are products most commonly distributed for use in the retail diesel fuel marketplace. Much of the world uses a system known as the "B" factor to state the amount of biodiesel in any fuel mix:

- 100% biodiesel is referred to as B100.
- 20% biodiesel, 80% petrodiesel is labeled B20.
- 5% biodiesel, 95% petrodiesel is labeled B5.
- 2% biodiesel, 98% petrodiesel is labeled B2.

Biodiesel sample.

Blends of 20% biodiesel and lower can be used in diesel equipment with no, or only minor modifications, although certain manufacturers do not extend warranty coverage if equipment is damaged by these blends. The B6 to B20 blends are covered by the ASTM D7467 specification. Biodiesel can also be used in its pure form (B100), but may require certain engine modifications to avoid maintenance and performance problems. Blending B100 with petroleum diesel may be accomplished by:

- Mixing in tanks at manufacturing point prior to delivery to tanker truck.
- Splash mixing in the tanker truck (adding specific percentages of biodiesel and petroleum diesel).
- In-line mixing, two components arrive at tanker truck simultaneously.
- Metered pump mixing, petroleum diesel and biodiesel meters are set to X total volume, transfer pump pulls from two points and mix is complete on leaving pump.

## Applications

Biodiesel can be used in pure form (B100) or may be blended with petroleum diesel at any concentration in most injection pump diesel engines. New extreme high-pressure (29,000 psi) common rail engines have strict factory limits of B5 or B20, depending on manufacturer. Biodiesel has different solvent properties from petrodiesel, and will degrade natural rubber gaskets and hoses in vehicles, although these tend to wear out naturally and most likely will have already been replaced with FKM, which is nonreactive to biodiesel. Biodiesel has been known to break down deposits of residue in the fuel lines where petrodiesel has been used. As a result, fuel filters may become clogged with particulates if a quick transition to pure biodiesel is made. Therefore, it is recommended to change the fuel filters on engines and heaters shortly after first switching to a biodiesel blend.

Targray Biofuels railcar transporting Biodiesel.

Since the passage of the Energy Policy Act of 2005, biodiesel use has been increasing in the United States. In the UK, the Renewable Transport Fuel Obligation obliges suppliers to include 5% renewable fuel in all transport fuel sold in the UK by 2010. For road diesel, this effectively means 5% biodiesel (B5).

## Vehicular use and Manufacturer Acceptance

In 2005, Chrysler (then part of DaimlerChrysler) released the Jeep Liberty CRD diesels from the factory into the European market with 5% biodiesel blends, indicating at least partial acceptance of biodiesel as an acceptable diesel fuel additive. In 2007, DaimlerChrysler indicated its intention to increase warranty coverage to 20% biodiesel blends if biofuel quality in the United States can be standardized.

The Volkswagen Group has released a statement indicating that several of its vehicles are compatible with B5 and B100 made from rape seed oil and compatible with the EN 14214 standard. The use of the specified biodiesel type in its cars will not void any warranty.

Mercedes Benz does not allow diesel fuels containing greater than 5% biodiesel (B5) due to concerns about "production shortcomings". Any damages caused by the use of such non-approved fuels will not be covered by the Mercedes-Benz Limited Warranty.

Starting in 2004, the city of Halifax, Nova Scotia decided to update its bus system to allow the fleet of city buses to run entirely on a fish-oil based biodiesel. This caused the city some initial mechanical issues, but after several years of refining, the entire fleet had successfully been converted.

In 2007, McDonald's of UK announced it would start producing biodiesel from the waste oil byproduct of its restaurants. This fuel would be used to run its fleet.

The 2014 Chevy Cruze Clean Turbo Diesel, direct from the factory, will be rated for up to B20 (blend of 20% biodiesel/80% regular diesel) biodiesel compatibility

## Railway Usage

British train operating company Virgin Trains West Coast claimed to have run the UK's first "biodiesel train", when a Class 220 was converted to run on 80% petrodiesel and 20% biodiesel.

Biodiesel locomotive and its external fuel tank.

The British Royal Train on 15 September 2007 completed its first ever journey run on 100% biodiesel fuel supplied by Green Fuels Ltd. Prince Charles and Green Fuels managing director James Hygate were the first passengers on a train fueled entirely by biodiesel fuel. Since 2007, the Royal Train has operated successfully on B100 (100% biodiesel).

Similarly, a state-owned short-line railroad in Eastern Washington ran a test of a 25% biodiesel/75% petrodiesel blend during the summer of 2008, purchasing fuel from a biodiesel producer sited along the railroad tracks. The train will be powered by biodiesel made in part from canola grown in agricultural regions through which the short line runs.

Also in 2007, Disneyland began running the park trains on B98 (98% biodiesel). The program was discontinued in 2008 due to storage issues, but in January 2009, it was announced that the park would then be running all trains on biodiesel manufactured from its own used cooking oils. This is a change from running the trains on soy-based biodiesel.

In 2007, the historic Mt. Washington Cog Railway added the first biodiesel locomotive to its all-steam locomotive fleet. The fleet has climbed up the western slopes of Mount Washington in New Hampshire since 1868 with a peak vertical climb of 37.4 degrees.

On 8 July 2014, the then Indian Railway Minister D.V. Sadananda Gowda announced in Railway Budget that 5% bio-diesel will be used in Indian Railways' Diesel Engines.

## Aircraft Use

A test flight has been performed by a Czech jet aircraft completely powered on biodiesel. Other recent jet flights using biofuel, however, have been using other types of renewable fuels.

On November 7, 2011 United Airlines flew the world's first commercial aviation flight on a microbially derived biofuel using Solajet, Solazyme's algae-derived renewable jet fuel. The Eco-skies Boeing 737-800 plane was fueled with 40 percent Solajet and 60 percent petroleum-derived jet fuel. The commercial Eco-skies flight 1403 departed from Houston's IAH airport at 10:30 and landed at Chicago's ORD airport at 13:03.

In September 2016, the Dutch flag carrier KLM contracted AltAir Fuels to supply all KLM flights departing Los Angeles International Airport with biofuel. For the next three years, the Paramount, California-based company will pump biofuel directly to the airport from their nearby refinery.

## Heating Oil

Biodiesel can also be used as a heating fuel in domestic and commercial boilers, a mix of heating oil and biofuel which is standardized and taxed slightly differently from diesel fuel used for transportation. Bioheat fuel is a proprietary blend of biodiesel and traditional heating oil. Bioheat is a registered trademark of the National Biodiesel Board [NBB] and the National Oilheat Research Alliance [NORA] in the U.S., and Columbia Fuels in Canada. Heating biodiesel is available in various blends. ASTM 396 recognizes blends of up to 5 percent biodiesel as equivalent to pure petroleum heating oil. Blends of higher levels of up to 20% biofuel are used by many consumers. Research is underway to determine whether such blends affect performance.

Older furnaces may contain rubber parts that would be affected by biodiesel's solvent properties, but can otherwise burn biodiesel without any conversion required. Care must be taken, however, given that varnishes left behind by petrodiesel will be released and can clog pipes- fuel filtering and prompt filter replacement is required. Another approach is to start using biodiesel as a blend, and decreasing the petroleum proportion over time can allow the varnishes to come off more gradually and be less likely to clog. Thanks to its strong solvent properties, however, the furnace is cleaned out and generally becomes more efficient. A technical research paper describes laboratory research and field trials project using pure biodiesel and biodiesel blends as a heating fuel in oil-fired boilers. During the Biodiesel Expo

2006 in the UK, Andrew J. Robertson presented his biodiesel heating oil research from his technical paper and suggested B20 biodiesel could reduce UK household $CO_2$ emissions by 1.5 million tons per year.

A law passed under Massachusetts Governor Deval Patrick requires all home heating diesel in that state to be 2% biofuel by July 1, 2010, and 5% biofuel by 2013. New York City has passed a similar law.

## Cleaning Oil Spills

With 80–90% of oil spill costs invested in shoreline cleanup, there is a search for more efficient and cost-effective methods to extract oil spills from the shorelines. Biodiesel has displayed its capacity to significantly dissolve crude oil, depending on the source of the fatty acids. In a laboratory setting, oiled sediments that simulated polluted shorelines were sprayed with a single coat of biodiesel and exposed to simulated tides. Biodiesel is an effective solvent to oil due to its methyl ester component, which considerably lowers the viscosity of the crude oil. Additionally, it has a higher buoyancy than crude oil, which later aids in its removal. As a result, 80% of oil was removed from cobble and fine sand, 50% in coarse sand, and 30% in gravel. Once the oil is liberated from the shoreline, the oil-biodiesel mixture is manually removed from the water surface with skimmers. Any remaining mixture is easily broken down due to the high biodegradability of biodiesel, and the increased surface area exposure of the mixture.

## Biodiesel in Generators

Biodiesel is also used in rental generators.

In 2001, UC Riverside installed a 6-megawatt backup power system that is entirely fueled by biodiesel. Backup diesel-fueled generators allow companies to avoid damaging blackouts of critical operations at the expense of high pollution and emission rates. By using B100, these generators were able to essentially eliminate the byproducts that result in smog, ozone, and sulfur emissions. The use of these generators in residential areas around schools, hospitals, and the general public result in substantial reductions in poisonous carbon monoxide and particulate matter.

## Properties

Biodiesel has promising lubricating properties and cetane ratings compared to low sulfur diesel fuels. Fuels with higher lubricity may increase the usable life of high-pressure fuel injection equipment that relies on the fuel for its lubrication. Depending on the engine, this might include high pressure injection pumps, pump injectors (also called unit injectors) and fuel injectors.

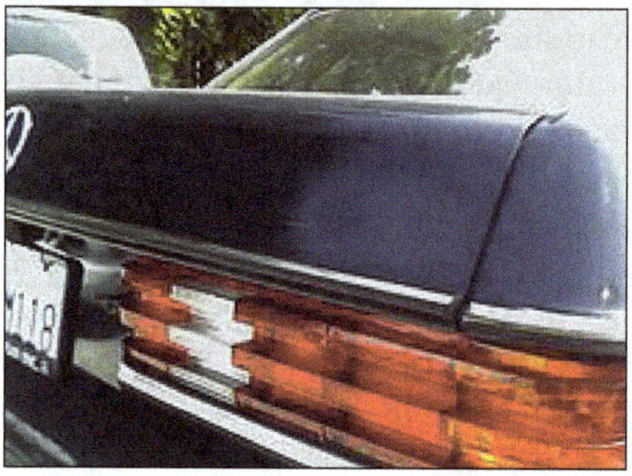

Older diesel Mercedes are popular for running on biodiesel.

The calorific value of biodiesel is about 37.27 MJ/kg. This is 9% lower than regular Number 2 petrodiesel. Variations in biodiesel energy density is more dependent on the feedstock used than the production process. Still, these variations are less than for petrodiesel. It has been claimed biodiesel gives better lubricity and more complete combustion thus increasing the engine energy output and partially compensating for the higher energy density of petrodiesel.

The color of biodiesel ranges from golden to dark brown, depending on the production method. It is slightly miscible with water, has a high boiling point and low vapor pressure. The flash point of biodiesel exceeds 130 °C (266 °F), significantly higher than that of petroleum diesel which may be as low as 52 °C (126 °F). Biodiesel has a density of ~0.88 g/cm$^3$, higher than petrodiesel (~0.85 g/cm$^3$).

Biodiesel contains virtually no sulfur, and it is often used as an additive to ultra-low-sulfur diesel (ULSD) fuel to aid with lubrication, as the sulfur compounds in petrodiesel provide much of the lubricity.

## Fuel Efficiency

The power output of biodiesel depends on its blend, quality, and load conditions under which the fuel is burnt. The thermal efficiency for example of B100 as compared to B20 will vary due to the differing energy content of the various blends. Thermal efficiency of a fuel is based in part on fuel characteristics such as: viscosity, specific density, and flash point; these characteristics will change as the blends as well as the quality of biodiesel varies. The American Society for Testing and Materials has set standards in order to judge the quality of a given fuel sample.

One study found that the brake thermal efficiency of B40 was superior to traditional petroleum counterpart at higher compression ratios (this higher brake thermal efficiency was recorded at compression ratios of 21:1). It was noted that, as the compression ratios increased, the efficiency of all fuel types – as well as blends being tested – increased; though it was found that a blend of B40 was the most economical at a compression ratio of 21:1 over all other blends. The study implied that this increase in efficiency was due to fuel density, viscosity, and heating values of the fuels.

## Combustion

Fuel systems on some modern diesel engines were not designed to accommodate biodiesel, while many heavy duty engines are able to run with biodiesel blends up to B20. Traditional direct injection fuel systems operate at roughly 3,000 psi at the injector tip while the modern common rail fuel system operates upwards of 30,000 PSI at the injector tip. Components are designed to operate at a great temperature range, from below freezing to over 1,000 °F (560 °C). Diesel fuel is expected to burn efficiently and produce as few emissions as possible. As emission standards are being introduced to diesel engines the need to control harmful emissions is being designed into the parameters of diesel engine fuel systems. The traditional inline injection system is more forgiving to poorer quality fuels as opposed to the common rail fuel system. The higher pressures and tighter tolerances of the common rail system allows for greater control over atomization and injection timing. This control of atomization as well as combustion allows for greater efficiency of modern diesel engines as well as greater control over emissions. Components within a diesel fuel system interact with the fuel in a way to ensure efficient operation of the fuel system and so the engine. If an out-of-specification fuel is introduced to a system that has specific parameters of operation, then the integrity of the overall fuel system may be compromised. Some of these parameters such as spray pattern and atomization are directly related to injection timing.

One study found that during atomization, biodiesel and its blends produced droplets greater in diameter than the droplets produced by traditional petrodiesel. The smaller droplets were attributed to the lower viscosity and surface tension of traditional diesel fuel. It was found that droplets at the periphery of the spray pattern were larger in diameter than the droplets at the center. This was attributed to the faster pressure drop at the edge of the spray pattern; there was a proportional relationship between the droplet size and the distance from the injector tip. It was found that B100 had the greatest spray penetration, this was attributed to the greater density of B100. Having a greater droplet size can lead to inefficiencies in the combustion, increased emissions, and decreased horse power. In another study it was found that there is a short injection delay when injecting biodiesel. This injection delay was attributed to the greater viscosity of Biodiesel. It was noted that the higher viscosity and the greater cetane rating of biodiesel over traditional petrodiesel lead to poor atomization, as well as mixture penetration with air during the ignition delay period. Another study noted that this ignition delay may aid in a decrease of $No_x$ emission.

## Emissions

Emissions are inherent to the combustion of diesel fuels that are regulated by the U.S. Environmental Protection Agency (E.P.A.). As these emissions are a byproduct of the combustion process, in order to ensure E.P.A. compliance a fuel system must be capable of controlling the combustion of fuels as well as the mitigation of emissions. There are a number of new technologies being phased in to control the production of diesel emissions. The exhaust gas recirculation system, E.G.R., and the diesel particulate filter, D.P.F., are both designed to mitigate the production of harmful emissions.

A study performed by the Chonbuk National University concluded that a B30 biodiesel blend reduced carbon monoxide emissions by approximately 83% and particulate matter emissions by roughly 33%. $No_x$ emissions, however, were found to increase without the application of an E.G.R. system. The study also concluded that, with E.G.R, a B20 biodiesel blend considerably reduced the emissions of the engine. Additionally, analysis by the California Air Resources Board found that biodiesel had the lowest carbon emissions of the fuels tested, those being ultra-low-sulfur diesel, gasoline, corn-based ethanol, compressed natural gas, and five types of biodiesel from varying feedstocks. Their conclusions also showed great variance in carbon emissions of biodiesel based on the feedstock used. Of soy, tallow, canola, corn, and used cooking oil, soy showed the highest carbon emissions, while used cooking oil produced the lowest.

While studying the effect of biodiesel on diesel particulate filters, it was found that though the presence of sodium and potassium carbonates aided in the catalytic conversion of ash, as the diesel particulates are catalyzed, they may congregate inside the D.P.F. and so interfere with the clearances of the filter. This may cause the filter to clog and interfere with the regeneration process. In a study on the impact of E.G.R. rates with blends of jathropa biodiesel it was shown that there was a decrease in fuel efficiency and torque output due to the use of biodiesel on a diesel engine designed with an E.G.R. system. It was found that CO and $CO_2$ emissions increased with an increase in exhaust gas recirculation but $No_x$ levels decreased. The opacity level of the jathropa blends was in an acceptable range, where traditional diesel was out of acceptable standards. It was shown that a decrease in $No_x$ emissions could be obtained with an E.G.R. system. This study showed an advantage over traditional diesel within a certain operating range of the E.G.R. system.

As of 2017, blended biodiesel fuels (especially B5, B8, and B20) are regularly used in many heavy-duty vehicles, especially transit buses in US cities. Characterization of exhaust emissions showed significant emission reductions compared to regular diesel.

## Material Compatibility

- Plastics: High-density polyethylene (HDPE) is compatible but polyvinyl chloride (PVC) is slowly degraded. Polystyrene is dissolved on contact with biodiesel.
- Metals: Biodiesel (like methanol) has an effect on copper-based materials (e.g. brass), and it also affects zinc, tin, lead, and cast iron. Stainless steels (316 and 304) and aluminum are unaffected.
- Rubber: Biodiesel also affects types of natural rubbers found in some older engine components. Studies have also found that fluorinated elastomers (FKM) cured with peroxide and base-metal oxides can be degraded when biodiesel loses its stability caused by oxidation. Commonly used synthetic rubbers FKM-GBL-S and FKM- GF-S found in modern vehicles were found to handle biodiesel in all conditions.

## Technical Standards

Biodiesel has a number of standards for its quality including European standard EN 14214, ASTM International D6751, and others.

## Low Temperature Gelling

When biodiesel is cooled below a certain point, some of the molecules aggregate and form crystals. The fuel starts to appear cloudy once the crystals become larger than one quarter of the wavelengths of visible light – this is the cloud point (CP). As the fuel is cooled further these crystals become larger. The lowest temperature at which fuel can pass through a 45 micrometre filter is the cold filter plugging point (CFPP). As biodiesel is cooled further it will gel and then solidify. Within Europe, there are differences in the CFPP requirements between countries. This is reflected in the different national standards of those countries. The temperature at which pure (B100) biodiesel starts to gel varies significantly and depends upon the mix of esters and therefore the feedstock oil used to produce the biodiesel. For example, biodiesel produced from low erucic acid varieties of canola seed (RME) starts to gel at approximately −10 °C (14 °F). Biodiesel produced from beef tallow and palm oil tends to gel at around 16 °C (61 °F) and 13 °C (55 °F) respectively. There are a number of commercially available additives that will significantly lower the pour point and cold filter plugging point of pure biodiesel. Winter operation is also possible by blending biodiesel with other fuel oils including #2 low sulfur diesel fuel and #1 diesel/kerosene.

Another approach to facilitate the use of biodiesel in cold conditions is by employing a second fuel tank for biodiesel in addition to the standard diesel fuel tank. The second fuel tank can be insulated and a heating coil using engine coolant is run through the tank. The fuel tanks can be switched over when the fuel is sufficiently warm. A similar method can be used to operate diesel vehicles using straight vegetable oil.

## Contamination by Water

Biodiesel may contain small but problematic quantities of water. Although it is only slightly miscible with water it is hygroscopic. One of the reasons biodiesel can absorb water is the persistence of mono and diglycerides left over from an incomplete reaction. These molecules can act as an emulsifier, allowing water to mix with the biodiesel. In addition, there may be water that is residual to processing or resulting from storage tank condensation. The presence of water is a problem because:

- Water reduces the heat of fuel combustion, causing smoke, harder starting, and reduced power.
- Water causes corrosion of fuel system components (pumps, fuel lines, etc.)
- Microbes in water cause the paper-element filters in the system to rot and fail, causing failure of the fuel pump due to ingestion of large particles.
- Water freezes to form ice crystals that provide sites for nucleation, accelerating gelling of the fuel.
- Water causes pitting in pistons.

Previously, the amount of water contaminating biodiesel has been difficult to measure by taking samples, since water and oil separate. However, it is now possible to measure the water content using water-in-oil sensors.

Water contamination is also a potential problem when using certain chemical catalysts involved in the production process, substantially reducing catalytic efficiency of base (high pH) catalysts such as potassium hydroxide. However, the super-critical methanol production methodology, whereby the transesterification process of oil feedstock and methanol is effectuated under high temperature and pressure, has been shown to be largely unaffected by the presence of water contamination during the production phase.

## Availability and Prices

Global biodiesel production reached 3.8 million tons in 2005. Approximately 85% of biodiesel production came from the European Union.

In 2007, in the United States, average retail (at the pump) prices, including federal and state fuel taxes, of B2/B5 were lower than petroleum diesel by about 12 cents, and

B20 blends were the same as petrodiesel. However, as part of a dramatic shift in diesel pricing, by July 2009, the US DOE was reporting average costs of B20 15 cents per gallon higher than petroleum diesel ($2.69/gal vs. $2.54/gal). B99 and B100 generally cost more than petrodiesel except where local governments provide a tax incentive or subsidy. In the month of October 2016, Biodiesel (B20) was 2 cents lower/gallon than petrodiesel.

In some countries biodiesel is less expensive than conventional diesel.

## Production

Biodiesel is commonly produced by the transesterification of the vegetable oil or animal fat feedstock, and other non-edible raw materials such as frying oil, etc. There are several methods for carrying out this transesterification reaction including the common batch process, heterogeneous catalysts, supercritical processes, ultrasonic methods, and even microwave methods.

Chemically, transesterified biodiesel comprises a mix of mono-alkyl esters of long chain fatty acids. The most common form uses methanol (converted to sodium methoxide) to produce methyl esters (commonly referred to as Fatty Acid Methyl Ester – FAME) as it is the cheapest alcohol available, though ethanol can be used to produce an ethyl ester (commonly referred to as Fatty Acid Ethyl Ester – FAEE) biodiesel and higher alcohols such as isopropanol and butanol have also been used. Using alcohols of higher molecular weights improves the cold flow properties of the resulting ester, at the cost of a less efficient transesterification reaction. A lipid transesterification production process is used to convert the base oil to the desired esters. Any free fatty acids (FFAs) in the base oil are either converted to soap and removed from the process, or they are esterified (yielding more biodiesel) using an acidic catalyst. After this processing, unlike straight vegetable oil, biodiesel has combustion properties very similar to those of petroleum diesel, and can replace it in most current uses.

The methanol used in most biodiesel production processes is made using fossil fuel inputs. However, there are sources of renewable methanol made using carbon dioxide or biomass as feedstock, making their production processes free of fossil fuels.

A by-product of the transesterification process is the production of glycerol. For every 1 tonne of biodiesel that is manufactured, 100 kg of glycerol are produced. Originally, there was a valuable market for the glycerol, which assisted the economics of the process as a whole. However, with the increase in global biodiesel production, the market price for this crude glycerol (containing 20% water and catalyst residues) has crashed. Research is being conducted globally to use this glycerol as a chemical building block. One initiative in the UK is The Glycerol Challenge.

Usually this crude glycerol has to be purified, typically by performing vacuum distillation. This is rather energy intensive. The refined glycerol (98%+ purity) can then be utilised directly, or converted into other products. The following announcements were made in 2007: A joint venture of Ashland Inc. and Cargill announced plans to make propylene glycol in Europe from glycerol and Dow Chemical announced similar plans for North America. Dow also plans to build a plant in China to make epichlorhydrin from glycerol. Epichlorhydrin is a raw material for epoxy resins.

## Production Levels

In 2007, biodiesel production capacity was growing rapidly, with an average annual growth rate from 2002–06 of over 40%. For the year 2006, the latest for which actual production figures could be obtained, total world biodiesel production was about 5–6 million tonnes, with 4.9 million tonnes processed in Europe (of which 2.7 million tonnes was from Germany) and most of the rest from the USA. In 2008 production in Europe alone had risen to 7.8 million tonnes. In July 2009, a duty was added to American imported biodiesel in the European Union in order to balance the competition from European, especially German producers. The capacity for 2008 in Europe totalled 16 million tonnes. This compares with a total demand for diesel in the US and Europe of approximately 490 million tonnes (147 billion gallons). Total world production of vegetable oil for all purposes in 2005/06 was about 110 million tonnes, with about 34 million tonnes each of palm oil and soybean oil. As of 2018, Indonesia is the world's top supplier of palmoil-based biofuel with annual production of 3.5 million tons, and expected to export about 1 million tonnes of biodiesel.

US biodiesel production in 2011 brought the industry to a new milestone. Under the EPA Renewable Fuel Standard, targets have been implemented for the biodiesel production plants in order to monitor and document production levels in comparison to total demand. According to the year-end data released by the EPA, biodiesel production in 2011 reached more than 1 billion gallons. This production number far exceeded the 800 million gallon target set by the EPA. The projected production for 2020 is nearly 12 billion gallons.

## Biodiesel Feedstocks

A variety of oils can be used to produce biodiesel. These include:

- Virgin oil feedstock – rapeseed and soybean oils are most commonly used, soybean oil accounting for about half of U.S. production. It also can be obtained from Pongamia, field pennycress and jatropha and other crops such as mustard, jojoba, flax, sunflower, palm oil, coconut and hemp.
- Waste vegetable oil (WVO).
- Animal fats including tallow, lard, yellow grease, chicken fat, and the by-products of the production of Omega-3 fatty acids from fish oil.
- Algae, which can be grown using waste materials such as sewage and without displacing land currently used for food production.
- Oil from halophytes such as Salicornia bigelovii, which can be grown using saltwater in coastal areas where conventional crops cannot be grown, with yields equal to the yields of soybeans and other oilseeds grown using freshwater irrigation.
- Sewage Sludge – The sewage-to-biofuel field is attracting interest from major companies like Waste Management and startups like InfoSpi, which are betting that renewable sewage biodiesel can become competitive with petroleum diesel on price.

Many advocates suggest that waste vegetable oil is the best source of oil to produce biodiesel, but since the available supply is drastically less than the amount of petroleum-based fuel that is burned for transportation and home heating in the world, this local solution could not scale to the current rate of consumption.

Animal fats are a by-product of meat production and cooking. Although it would not be efficient to raise animals (or catch fish) simply for their fat, use of the by-product adds value to the livestock industry (hogs, cattle, poultry). Today, multi-feedstock biodiesel facilities are producing high quality animal-fat based biodiesel. Currently, a 5-million dollar plant is being built in the US, with the intent of producing 11.4 million litres (3 million gallons) biodiesel from some of the estimated 1 billion kg (2.2 billion pounds) of chicken fat produced annually at the local Tyson poultry plant. Similarly, some small-scale biodiesel factories use waste fish oil as feedstock. An EU-funded project (ENERFISH) suggests that at a Vietnamese plant to produce biodiesel from catfish (basa, also known as pangasius), an output of 13 tons/day of biodiesel can be produced from 81 tons of fish waste (in turn resulting from 130 tons of fish). This project utilises the biodiesel to fuel a CHP unit in the fish processing plant, mainly to power the fish freezing plant.

## Quantity of Feedstocks Required

Current worldwide production of vegetable oil and animal fat is not sufficient to replace liquid fossil fuel use. Furthermore, some object to the vast amount of farming and the resulting fertilization, pesticide use, and land use conversion that would be needed to produce the additional vegetable oil. The estimated transportation diesel fuel and home heating oil used in the United States is about 160 million tons (350 billion pounds) according to the Energy Information Administration, US Department of Energy. In the United States, estimated production of vegetable oil for all uses is about 11 million tons (24 billion pounds) and estimated production of animal fat is 5.3 million tonnes (12 billion pounds).

If the entire arable land area of the USA (470 million acres, or 1.9 million square kilometers) were devoted to biodiesel production from soy, this would just about provide the 160 million tonnes required (assuming an optimistic 98 US gal/acre of biodiesel). This land area could in principle be reduced significantly using algae, if the obstacles can be overcome. The US DOE estimates that if algae fuel replaced all the petroleum fuel in the United States, it would require 15,000 square miles (39,000 square kilometers), which is a few thousand square miles larger than Maryland, or 30% greater than the area of Belgium, assuming a yield of 140 tonnes/hectare (15,000 US gal/acre). Given a more realistic yield of 36 tonnes/hectare (3834 US gal/acre) the area required is about 152,000 square kilometers, or roughly equal to that of the state of Georgia or of England and Wales. The advantages of algae are that it can be grown on non-arable land such as deserts or in marine environments, and the potential oil yields are much higher than from plants.

## Yield

Feedstock yield efficiency per unit area affects the feasibility of ramping up production to the huge industrial levels required to power a significant percentage of vehicles.

Algae fuel yields have not yet been accurately determined, but DOE is reported as saying that algae yield 30 times more energy per acre than land crops such as soybeans. Yields of 36 tonnes/hectare are considered practical by Ami Ben-Amotz of the Institute of Oceanography in Haifa, who has been farming Algae commercially for over 20 years.

Jatropha has been cited as a high-yield source of biodiesel but yields are highly dependent on climatic and soil conditions. The estimates at the low end put the yield at about 200 US gal/acre (1.5-2 tonnes per hectare) per crop; in more favorable climates two or more crops per year have been achieved. It is grown in the Philippines, Mali and India, is drought-resistant, and can share space with other cash crops such as coffee, sugar, fruits and vegetables. It is well-suited to semi-arid lands and can contribute to slow down desertification, according to its advocates.

## Efficiency and Economic Arguments

According to a study by Drs. Van Dyne and Raymer for the Tennessee Valley Authority, the average US farm consumes fuel at the rate of 82 litres per hectare (8.75 US gal/acre) of land to produce one crop. However, average crops of rapeseed produce oil at an average rate of 1,029 L/ha (110 US gal/acre), and high-yield rapeseed fields produce about 1,356 L/ha (145 US gal/acre). The ratio of input to output in these cases is roughly 1:12.5 and 1:16.5. Photosynthesis is known to have an efficiency rate of about 3–6% of total solar radiation and if the entire mass of a crop is utilized for energy production, the overall efficiency of this chain is currently about 1% While this may compare unfavorably to solar cells combined with an electric drive train, biodiesel is less costly to deploy (solar cells cost approximately US$250 per square meter) and transport (electric vehicles require batteries which currently have a much lower energy density than liquid fuels). A 2005 study found that biodiesel production using soybeans required 27% more fossil energy than the biodiesel produced and 118% more energy using sunflowers.

Pure biodiesel (B-100) made from soybeans.

However, these statistics by themselves are not enough to show whether such a change makes economic sense. Additional factors must be taken into account, such as: the fuel equivalent of the energy required for processing, the yield of fuel from raw oil, the return on cultivating food, the effect biodiesel will have on food prices and the relative cost of biodiesel versus petrodiesel, water pollution from farm run-off, soil depletion, and the externalized costs of political and military interference in oil-producing countries intended to control the price of petrodiesel.

The debate over the energy balance of biodiesel is ongoing. Transitioning fully to biofuels could require immense tracts of land if traditional food crops are used (although non food crops can be utilized). The problem would be especially severe for nations with large economies, since energy consumption scales with economic output.

If using only traditional food plants, most such nations do not have sufficient arable land to produce biofuel for the nation's vehicles. Nations with smaller economies (hence less energy consumption) and more arable land may be in better situations, although many regions cannot afford to divert land away from food production.

For third world countries, biodiesel sources that use marginal land could make more sense; e.g., pongam oiltree nuts grown along roads or jatropha grown along rail lines.

In tropical regions, such as Malaysia and Indonesia, plants that produce palm oil are being planted at a rapid pace to supply growing biodiesel demand in Europe and other markets. Scientists have shown that the removal of rainforest for palm plantations is not ecologically sound since the expansion of oil palm plantations poses a threat to natural rainforest and biodiversity.

It has been estimated in Germany that palm oil biodiesel has less than one third of the production costs of rapeseed biodiesel. The direct source of the energy content of biodiesel is solar energy captured by plants during photosynthesis. Regarding the positive energy balance of biodiesel.

When straw was left in the field, biodiesel production was strongly energy positive, yielding 1 GJ biodiesel for every 0.561 GJ of energy input (a yield/cost ratio of 1.78).

When straw was burned as fuel and oilseed rapemeal was used as a fertilizer, the yield/cost ratio for biodiesel production was even better (3.71). In other words, for every unit of energy input to produce biodiesel, the output was 3.71 units (the difference of 2.71 units would be from solar energy).

## Economic Impact

Multiple economic studies have been performed regarding the economic impact of biodiesel production. One study, commissioned by the National Biodiesel Board, reported the production of biodiesel supported more than 64,000 jobs. The growth in biodiesel also helps significantly increase GDP. In 2011, biodiesel created more than $3 billion in GDP. Judging by the continued growth in the Renewable Fuel Standard and the extension of the biodiesel tax incentive, the number of jobs can increase to 50,725, $2.7 billion in income, and reaching $5 billion in GDP by 2012 and 2013.

## Energy Security

One of the main drivers for adoption of biodiesel is energy security. This means that a nation's dependence on oil is reduced, and substituted with use of locally available sources, such as coal, gas, or renewable sources. Thus a country can benefit from adoption of biofuels, without a reduction in greenhouse gas emissions. While the total energy balance is debated, it is clear that the dependence on oil is reduced. One example is the energy used to manufacture fertilizers, which could come from a variety of sources other than petroleum. The US National Renewable Energy Laboratory (NREL) states that energy security is the number one driving force behind the US biofuels programme, and a White House "Energy Security for the 21st Century" paper makes it clear that energy security is a major reason for promoting biodiesel. The former EU commission president, Jose Manuel Barroso, speaking at a recent EU biofuels conference, stressed that properly managed biofuels have the potential to reinforce the EU's security of supply through diversification of energy sources.

## Environmental Effects

The surge of interest in biodiesels has highlighted a number of environmental effects associated with its use. These potentially include reductions in greenhouse gas emissions, deforestation, pollution and the rate of biodegradation.

According to the EPA's Renewable Fuel Standards Program Regulatory Impact Analysis, released in February 2010, biodiesel from soy oil results, on average, in a 57% reduction in greenhouse gases compared to petroleum diesel, and biodiesel produced from waste grease results in an 86% reduction.

However, environmental organizations, for example, Rainforest Rescue and Greenpeace, criticize the cultivation of plants used for biodiesel production, e.g., oil palms, soybeans and sugar cane. They say the deforestation of rainforests exacerbates climate change and that sensitive ecosystems are destroyed to clear land for oil palm, soybean and sugar cane plantations. Moreover, that biofuels contribute to world hunger, seeing as arable land is no longer used for growing foods. The Environmental Protection Agency (EPA) published data in January 2012, showing that biofuels made from palm oil will not count towards the nation's renewable fuels mandate as they are not climate-friendly. Environmentalists welcome the conclusion because the growth of oil palm plantations has driven tropical deforestation, for example, in Indonesia and Malaysia.

## Food, Land and Water vs. Fuel

In some poor countries the rising price of vegetable oil is causing problems. Some propose that fuel only be made from non-edible vegetable oils such as camelina, jatropha or seashore mallow which can thrive on marginal agricultural land where many trees and crops will not grow, or would produce only low yields.

Others argue that the problem is more fundamental. Farmers may switch from producing food crops to producing biofuel crops to make more money, even if the new crops are not edible. The law of supply and demand predicts that if fewer farmers are producing food the price of food will rise. It may take some time, as farmers can take some time to change which things they are growing, but increasing demand for first generation biofuels is likely to result in price increases for many kinds of food. Some have pointed out that there are poor farmers and poor countries who are making more money because of the higher price of vegetable oil.

Biodiesel from sea algae would not necessarily displace terrestrial land currently used for food production and new algaculture jobs could be created.

By comparison it should be mentioned that the production of biogas utilizes agricultural waste to generate a biofuel known as biogas, and also produces compost, thereby enhancing agriculture, sustainability and food production.

## Concerns

### Engine Wear

Lubricity of fuel plays an important role in wear that occurs in an engine. A diesel engine relies on its fuel to provide lubricity for the metal components that are constantly in contact with each other. Biodiesel is a much better lubricant compared with petroleum diesel due to the presence of esters. Tests have shown that the addition of a small amount of biodiesel to diesel can significantly increase the lubricity of the fuel in short term. However, over a longer period of time (2–4 years), studies show that biodiesel loses its lubricity. This could be because of enhanced corrosion over time due to oxidation of the unsaturated molecules or increased water content in biodiesel from moisture absorption.

### Fuel Viscosity

One of the main concerns regarding biodiesel is its viscosity. The viscosity of diesel is 2.5–3.2 cSt at 40 °C and the viscosity of biodiesel made from soybean oil is between 4.2 and 4.6 cSt The viscosity of diesel must be high enough to provide sufficient lubrication for the engine parts but low enough to flow at operational temperature. High viscosity can plug the fuel filter and injection system in engines. Vegetable oil is composed of lipids with long chains of hydrocarbons, to reduce its viscosity the lipids are broken down into smaller molecules of esters. This is done by converting vegetable oil and animal fats into alkyl esters using transesterification to reduce their viscosity Nevertheless, biodiesel viscosity remains higher than that of diesel, and the engine may not be able to use the fuel at low temperatures due to the slow flow through the fuel filter.

## References


- What-is-bioenergy-2941107: thebalance.com, Retrieved 12 May, 2020
- Munnings, C.; Kulkarni, A.; Giddey, S.; Badwal, S.P.S. (August 2014). "Biomass to power conversion in a direct carbon fuel cell". International Journal of Hydrogen Energy. 39 (23): 12377–12385. doi:10.1016/j.ijhydene.2014.03.255
- Agostini, Francesco; Gregory, Andrew S.; Richter, Goetz M. (15 January 2015). "Carbon Sequestration by Perennial Energy Crops: Is the Jury Still Out?". BioEnergy Research. 8 (3): 1057–1080. doi:10.1007/s12155-014-9571-0. PMC 4732603. PMID 26855689
- Vnokurov, V. A.; A.V. Barkov; L. M. Krasnopol'skya; E.S. Mortikov (2 November 2010). "CURRENT PROBLEMS. Alternative Fuels Technology". Chemistry and Technology of Fuels and Oils. 46 (2): 75–78. doi:10.1007/s10553-010-0190-y
- Read, Peter; Lermit, Jonathan (2005). "Bio-energy with carbon storage (BECS): A sequential decision approach to the threat of abrupt climate change". Energy. 30 (14): 2654. doi:10.1016/j.energy.2004.07.003
- Qureshi, Nasib; Hodge, David; Vertès, Alain (2014). Biorefineries. Integrated Biochemical Processes for Liquid Biofuels. Elsevier. p. 59. ISBN 9780444594983
- 10-disadvantages-of-biofuels, fuel-efficiency-biofuels: auto.howstuffworks.com, Retrieved 13 July, 2020

# PERMISSIONS

All chapters in this book are published with permission under the Creative Commons Attribution Share Alike License or equivalent. Every chapter published in this book has been scrutinized by our experts. Their significance has been extensively debated. The topics covered herein carry significant information for a comprehensive understanding. They may even be implemented as practical applications or may be referred to as a beginning point for further studies.

We would like to thank the editorial team for lending their expertise to make the book truly unique. They have played a crucial role in the development of this book. Without their invaluable contributions this book wouldn't have been possible. They have made vital efforts to compile up to date information on the varied aspects of this subject to make this book a valuable addition to the collection of many professionals and students.

This book was conceptualized with the vision of imparting up-to-date and integrated information in this field. To ensure the same, a matchless editorial board was set up. Every individual on the board went through rigorous rounds of assessment to prove their worth. After which they invested a large part of their time researching and compiling the most relevant data for our readers.

The editorial board has been involved in producing this book since its inception. They have spent rigorous hours researching and exploring the diverse topics which have resulted in the successful publishing of this book. They have passed on their knowledge of decades through this book. To expedite this challenging task, the publisher supported the team at every step. A small team of assistant editors was also appointed to further simplify the editing procedure and attain best results for the readers.

Apart from the editorial board, the designing team has also invested a significant amount of their time in understanding the subject and creating the most relevant covers. They scrutinized every image to scout for the most suitable representation of the subject and create an appropriate cover for the book.

The publishing team has been an ardent support to the editorial, designing and production team. Their endless efforts to recruit the best for this project, has resulted in the accomplishment of this book. They are a veteran in the field of academics and their pool of knowledge is as vast as their experience in printing. Their expertise and guidance has proved useful at every step. Their uncompromising quality standards have made this book an exceptional effort. Their encouragement from time to time has been an inspiration for everyone.

The publisher and the editorial board hope that this book will prove to be a valuable piece of knowledge for students, practitioners and scholars across the globe.

# INDEX

www.ingramcontent.com/pod-product-compliance
Lightning Source LLC
LaVergne TN
LVHW080623210926
848485LV00003B/203